The Social Costs of
Solar Energy

Pergamon Titles of Related Interest

Grenon THE NUCLEAR APPLE AND THE SOLAR ORANGE
Murphy ENERGY AND THE ENVIRONMENTAL BALANCE
Secretariat for Future Studies SOLAR VERSUS NUCLEAR
Stewart TRANSITIONAL ENERGY POLICY 1980-2030
Ural ENERGY RESOURCES AND CONSERVATION RELATED TO
    BUILT ENVIRONMENT

Related Journals*

ENVIRONMENT INTERNATIONAL
THE ENVIRONMENTAL PROFESSIONAL
ENERGY, THE INTERNATIONAL JOURNAL
SOLAR ENERGY
SPACE SOLAR POWER REVIEW
SUN AT WORK IN BRITAIN
SUNWORLD

*Free specimen copies available upon request.

PERGAMON POLICY STUDIES

ON SCIENCE AND TECHNOLOGY

The Social Costs of Solar Energy
A Study of Photovoltaic Energy Systems

Thomas L. Neff

Pergamon Press
NEW YORK • OXFORD • TORONTO • SYDNEY • PARIS • FRANKFURT

Pergamon Press Offices:

U.S.A.   Pergamon Press Inc., Maxwell House, Fairview Park,
                         Elmsford, New York 10523, U.S.A.

U.K.   Pergamon Press Ltd., Headington Hill Hall,
                         Oxford OX3 OBW, England

CANADA   Pergamon of Canada, Ltd., Suite 104, 150 Consumers Road,
                         Willowdale, Ontario M2J 1P9, Canada

AUSTRALIA   Pergamon Press (Aust.) Pty. Ltd., P.O. Box 544,
                         Potts Point, NSW 2011, Australia

FRANCE   Pergamon Press SARL, 24 rue des Ecoles,
                         75240 Paris, Cedex 05, France

FEDERAL REPUBLIC   Pergamon Press GmbH, Hammerweg 6, Postfach 1305,
OF GERMANY   6242 Kronberg/Taunus, Federal Republic of Germany

Library of Congress Cataloging in Publication Data

Neff, Thomas L
  The social costs of solar energy.

  (Pergamon policy studies on science and technology)
  1. Solar energy—Social aspects. I. Title.
II. Series.
HD9681.A2N43 1981   338.4'762147   80-23732
ISBN 0-08-026315-1

Work prepared pursuant to author's contract
with U.S. Department of Energy who retain
non-exclusive, irrevocable license to reproduce,
publish, use and dispose of all material
contained herein.

To
Chris and Marc

Contents

Preface & Acknowledgments

Solar energy is often regarded as a totally clean and hazard-free alternative to conventional energy sources that have numerous negative impacts on health, the environment, political and social processes, or on domestic and international security. This view of solar energy is, of course, born primarily of hope. Indeed, it is unlikely that any energy technology—whatever its social benefits—will be entirely free of social costs. This study makes a first effort to provide a comparative analytical basis for social perspectives on one important solar technology, the direct photovoltaic conversion of sunlight to electricity.

Several students and other individuals contributed substantially to this study: James Durand to chapter 5 and the appendix; John Katrakis to chapters 4 and 7; Subramanyam Kumar and John Cochrane to chapter 7; and Virginia Faust to the collection of data and the overall organization of the report. Special thanks are due Professor S. Sieferman for editorial and other improvements subsequent to the completion of the draft report in May, 1978.

Chapter 1
Introduction

Social costs are associated with all commonly used electrical energy sources. These costs arise from effects on public and occupational health and on the environment; they also originate in a number of important effects on social and political processes. The use of coal, for example, not only affects the health of miners and the public and the environment (throughout such impacts as mine drainage, acid rain, and a possible increase in global CO_2 levels) but also raises other national and even international political and social issues. Among these are conflicts resulting from power plant siting, from inequities between coal-producing and coal-consuming regions, and from the movement of pollutants from coal combustion across state or national boundaries. For nuclear power the mix of such impacts is different, but significant impacts certainly exist. Thus, for example, while nuclear power probably involves lower average impacts on health and environment than coal under normal operating conditions, there are countervailing negative effects from the possibility of nuclear accidents, from the difficulties in managing the disposal of wastes, and from the international security threat of nuclear weapons proliferation. The impacts associated with conventional electrical energy sources, despite decades of use, are still highly uncertain in their identity and magnitude; experience suggests that we have not yet become aware of the extent to which even conventional energy technologies involve significant social costs.

The existence of substantial social costs (known and possible) associated with conventional technologies provides incentive to develop new energy technologies characterized by demonstrably small impacts on human health and the environment or, perhaps, by contributing less to the causes of social and political conflict than do conventional sources. In the popular mind, solar energy technologies have come to be associated with precisely these virtues. Among the solar technologies, thermal electric and photovoltaic generation of electricity are those most often thought of as substitutable for current electrical generation systems. While popular conceptions undoubtedly play a role in the initial acceptance of a new technology, current policy decisions, and the long-term

1

success of a new technology, must rest on critical evaluation of the social costs associated with that technology, as compared with conventional alternatives. Given large uncertainties about these alternatives, and the even larger uncertainties associated with a yet immature new technology, this is indeed difficult. It is not, however, impossible to make such evaluations, at least qualitatively, and in some instances, quantitatively.

In this study we shall present a preliminary social cost evaluation of photovoltaic technologies, emphasizing qualitative comparisons with electricity generation systems based on coal (and, in some cases, nuclear fuels).* Where possible and important in this study, quantitative assessments of photovoltaic impacts will be performed. These typically will be rough calculations intended to establish relative priorities in policy issues. In many cases, calculations will be performed in the spirit of upper bound—or worst case—situations. Calculations of this type are of great value if it can be shown that the *maximum* possible impacts of new technology are well below the *minimum* impacts known to be associated with conventional sources of electricity. The value of this type of comparison is especially great because the large uncertainties currently involved in evaluation of conventional technologies, especially coal, lead to a perceived (and very likely also a real) probability of increasingly high social costs. Thus, if it is possible to establish firm upper bounds on photovoltaic impacts, one has thereby established a basis for evaluating the value of that technology in avoiding not only the currently known costs of conventional sources, but also the risk of currently unknown, but potentially high, costs which might eventually be found to be associated with them. It should be remembered, however, that upper bound calculations probably overstate the true impacts of photovoltaics and should be used only for the policy purposes indicated. Careful and comprehensive health and environmental analyses will be required for dealing with more detailed policy issues and for establishing the regulatory basis on which photovoltaic devices will be manufactured and used.

Comparison of the impacts of different energy technologies can be examined in the several dimensions affected:

- Occupational safety and health impacts arising at the various stages of production, installation, and use. Both immediate effects such as accidents or toxic exposures and long-term health effects must be considered.
- Public health impacts arising from the above stages. These result from routine emissions or from accidents and generally involve assessment of long-term,

---

*Ultimately, photovoltaics may be competitive with coal and nuclear power; in the interim, it will more likely compete with oil and natural gas for peak load generation. The comparison in this case is somewhat more difficult since the primary social costs associated with oil and gas (except for sulfur and particulates for some oil) are opportunity costs: the drawing of oil and gas away from other uses and resulting price increases due to higher demand.

low-level health effects as well as immediate effects. The standards for public health effects are generally much more restrictive than those for occupational impacts.

- Environmental effects, including impacts on ecosystems, land use, and thermal, climatic and other effects.
- Impacts on social, political, economic, and institutional processes. Different energy devices involve different mixes of the factors of production: among them, capital, labor, energy, and materials. The applications of these technologies also have major implications for societal decisions and decision processes, for choices of living and development patterns, and for the balances between regional or ideological political interests. While the nature of these impacts may be described independently, relative assessment invariably involves social judgments.

In subsequent chapters, we shall deal with the first three of these categories, on a comparative basis, for photovoltaic technologies and applications. With respect to the fourth category, we limit our attention here to the effects attributable to differing factors of production. The broader sociopolitical issues remain to be addressed in future research.

Before embarking on our analysis, it is useful to identify some of the methodological problems which must be resolved in comparing photovoltaic systems with conventional and alternative energy sources and to sketch the important characteristics of photovoltaic technologies.

Chapter 2
Methodological Issues

In assessing the relative health and environmental impacts of photovoltaic and alternative technologies, several methodological problems arise. These include establishing the basis for proper allocation of social costs, the incomparabilities of different types of social costs and the impossibility of making value-free overall comparisons, constraints on the displacement of social costs associated with one energy technology by those of another, and the difficulty of accounting for social costs only indirectly related to the use of a given energy technology. Since the resolution of these problems strongly affects social cost comparisons, each deserves discussion in greater detail.

Social Cost Allocation

The first methodological issue is a normalization problem stemming from the fact that sunlight is an inconstant source* of energy. While fossil and nuclear facilities may be operated as continuously as is technically feasible, solar power installations are limited by variations in insolation (diurnal, seasonal and weather-induced variations). Because of this, social costs cannot be assessed on the basis of installed peak generating capacity. Instead, it is useful to prorate effects over the electrical energy produced; for example, the deaths or injuries or emissions levels ascribable to the generation of a kilowatt-hour or gigawatt-electric year GWe-yr) of electricity. The latter unit, equivalent to 8.76 to 10^9 kwh,[1] will be used throughout this study.**

While this device does a great deal to reduce very different technologies to a common measure of social disutility, it should be noted that it is not a perfect basis for a relative social cost/benefit comparison since a kilowatt-hour produced

---

*We do not consider earth satellite photovoltaic power systems in this study.

**Under average conditions, a coal or nuclear plant with a rated capacity of 1 GWe will produce only about 0.6 GWe-yr of electricity in a year; photovoltaic arrays with a comparable total peak output will produce about 0.2 GWe-yr in a year (the peak to average output ratio of 5 is commonly used, and is used here, but clearly varies with geographical area).

4

at the time of peak demand may be regarded as providing a higher social benefit. Since peak photovoltaic generation generally coincides with the onset of peak demand, it can be used to relieve peak load generating requirements (now generally satisfied by burning oil or natural gas), *if* it is used in the electric grid. However, if photovoltaics were to be used in stand-alone applications, it would be necessary to use energy storage systems, such as batteries or flywheels, to hold power until needed. These storage systems usually involve additional social costs which may or may not be equal to the additional benefit of having the power available on a continuous basis. For purposes of comparison in this study, we ignore these use parameters. However, they may be of great importance in determining the social utility of specific proposed installations.

Incomparability and Social Values

The second methodological problem in comparing energy technologies is that different technologies have qualitatively different impacts and thus may not be directly comparable. For example, it is difficult to compare the health impact of coal mining on coal miners with the effects on the public of perceptions of risks associated with nuclear wastes. Impacts may be different in character, may affect different groups or institutions in society, and may even affect different generations. In dealing with this problem, the first step in analysis is clearly to estimate, within the limits set by imperfect knowledge, social impacts in various categories—occupational deaths, injuries, delayed health effects, analogous public health impacts, and so forth. For each energy technology one then has a balance sheet with many columns, with the entries in each column expressed in a different currency of social disutility.

While it is then possible to compare technologies in each category of impact, it is not possible to make a direct overall comparison between the impacts of different energy technologies without imposing social judgments about the relative importance of various types of impacts (the currency exchange values, as it were). For example, only society can decide whether an occupational death is more or less important than a given number of cases of degradation of health of members of the public, or a given level of risk to future generations. Efforts have been made to find a common currency—usually an equivalent dollar value for lives and other effects—and to the extent that society has commonly expressed such judgments (as in the case of occupational hazards), such efforts provide some guidance. However, it is not possible to provide an overall comparison between technologies in a way which is free of the social values of the analyst.

This lack of commonality is especially important in considering choices between energy technologies which have very different characteristics, since preference for avoiding one particular type of social cost may lead to a judgment in favor of one technology while emphasis on another cost element may lead to the opposite. This difference in social focus is in part responsible for the controversy over the choice between coal and nuclear power, where a choice must be made between widespread, but perhaps individually modest, aggravation of public health in the case of coal, and low-probability/high-consequence accidents (involving reactors or, perhaps, nuclear wastes) in the case of nuclear power. This situation is further complicated by lack of information. If such problems can arise in the comparison of two technologies which otherwise have rather similar configurations, they will be even more profound for comparisons of conventional sources of electricity with photovoltaic systems, which may be quite different. This is especially important for the fourth class of impacts listed above: the effects on the social/economic/political/institutional system.

Substitution and Energy Interdependencies

The third methodological problem is that deployment of one energy technology must be seen as displacing use of another. Accounting for this raises the allocation and commonality issues above, and incomplete consideration again may give misleading results. There is, however, another important subtlety involved in the social cost accounting. Manufacture and deployment of new energy generation devices involves the use of energy from conventional sources. Use of this energy incurs a social cost debt for the new energy device, since there is directly attributable to deployment of the new devices some use of conventional sources which otherwise would not be used or would be used for other purposes. Even if a new technology entails no *direct* social costs, its use must therefore be seen as entailing energy-related indirect social costs.

These costs, which are characteristic of the mix of technologies in use at the time of deployment of the new energy generation device, should be prorated over the lifetime energy output of the new device (we ignore possible discounting of future social costs). Then the introduction of a new technology results, in effect, in a dilution of the social costs associated with pre-existing energy sources (if the new technology, in fact, involves lower direct social costs). The magnitude of this dilution depends on the energy investment from conventional sources and on the lifetime energy return on this investment. For example, if the new energy technology device (e.g., an installed solar panel) requires an investment (say of energy from coal) equal to five years' output, and has a life of 20 years, each unit of energy input is eventually transformed into four units of energy output, resulting in a reduction of average energy-related social costs by a factor of four, if the direct costs of the new technology are negligible. In the longer term, deployment of the new "clean" technology will gradually displace technologies which are socially more costly—itself a social benefit. But the extent and rate at which a new technology can bring an alleviation of social impacts of energy use in the near term is small.

Indirect Social Costs

The fourth problem that arises in the assessment of any energy (or other) technology is that there are important underlying and antecedent social costs that may be less evident than the direct effects of the technology. For example, the manufacture of energy devices involves the use of materials like aluminum or steel. The production of these materials may affect the environment, the health of workers, or the public. These underlying costs may in fact be larger than the direct impacts of the technology. Antecedent even to these underlying costs, there are also activities with social impacts. The construction of a steel mill to make the steel in our example above also involves social costs. The difficulty here is that there is no obvious limit on how far back to go, nor in how widely to range, in bookkeeping social costs. A truly complete bookkeeping of social costs associated with a given energy technology would require a comprehensive examination of the very fabric of society.

Fortunately, this problem may drop out in some comparisons between energy technologies. New energy technologies may place comparable burdens on the underlying productive organization of society; for example, the amounts of material and capital used in a nuclear plant (and its associated distribution grid) may be comparable to those required for a decentralized system of photovoltaic devices of similar total energy output. Where this is not the case, the most important effects can be accounted for largely through consideration of the pervasive factors of production: labor, materials, and energy.* It is this approach that we use in chapter 7 to account for indirect social costs.

REFERENCES

1. About 2×10^{12} Kwh, or 228 GWe-yr of electricity, were used in the U. S. in 1976.

---

*Capital is a fourth major factor but is not considered in this analysis.

Chapter 3
Potential Risk Centers in Photovoltaic Technologies

Photovoltaic technologies and production processes are far from mature, resulting in a lack of information on precisely how these technologies may be used and a lack of data on what impacts actually occur. Nevertheless, preliminary health and environmental analysis of generic technological choices can play an extremely important role in decisions made concerning photovoltaic technologies. Indeed, analysis of present processes may help in determining which suboptions are most attractive from a social cost standpoint, or which process steps need improvement before an option may be considered acceptable.

For purposes of analysis, this study will consider the three main lines of photovoltaic research and development effort.*

- large-crystal silicon (Si) wafer cells in flat plate arrays
- cadmium sulfide (CdS) cells in flat plate arrays
- gallium arsenide (GaAs) cells in flat plate or concentrator arrays

There are three categories of applications:

- decentralized residential installations
- decentralized neighborhood, commercial or industrial applications
- central station plants

Conventional and other advanced technologies can be compared with these systems and applications, especially at points of identified risk.**

---

*These technologies are currently at rather different stages of development; however, all are potentially viable candidates for future applications.

**To the extent that storage systems are necessary to shift energy from periods of peak insolation to periods of use, it is necessary to bookkeep the associated social costs. We shall not consider storage technologies in this study except to note that the manufacture and use of such systems are not without hazards.

8

It is necessary to characterize each technology in somewhat more detail before proceeding with an evaluation of occupational, public, and other impacts.*

Silicon

Silicon cells are currently manufactured using materials and processes developed in the semiconductor industry. Silicon dioxide is mined as sand or quartzite and must be reduced by heating with coke in an electric arc furnace to produce metallurgical-grade silicon of 96–98 percent purity. These steps are probably essential for any silicon-based PV technology. The amount of silicon reduced and refined per unit of installed capacity depends on cell thickness, photovoltaic efficiency and, especially, on the process efficiencies assumed.

Representative figures for silicon consumed are given in fig. 5.1 as 135 tonnes of SiO_2 mined per peak MWe of PV capacity, resulting ultimately in 2.5 tonnes/ MWe peak of semiconductor grade SiO_2 in PV arrays.** All of these figures depend on technological assumptions which may change in time. SiO or SiO_2 particulates and flyash from coke during electric arc refining are the principal source of direct occupational and public health concerns. The processes by which submicron particulates find their way to humans are explored in subsequent sections. Once inhaled, however, silicon particulates can result in scarring of lung tissue (silicosis) or be translocated to the kidneys or other sensitive organs. Synergistic effects involving other pollutants are also likely (e.g., particulates may provide catalytic surfaces for reactions involving chemical pollutants or a transport mechanism into the lung) and impairment of lung function may increase the likelihood of other diseases. The presence of SiO (which is more reactive than SiO_2) in emissions, and its role in health effects, are of particular interest.

Present cells require further refinement of metallurgical silicon to semiconductor grade (through chlorination, hydrogenation, reactive gas blow-through, zone refining, directional freezing, or combination of these). Process steps and chemicals used may involve hazards which could be studied in existing industries. The purified silicon is then formed into single crystals or polycrystalline ingots, by one of several processes, and cut into wafers or blanks for cells. New processes include ribbon, sheet, vapor deposition, and other processes producing a "solar-grade" material of varying crystalline character. Conventional processes may involve hazards from silicon dust or cutting oils.

Cell Fabrication and Array Assembly

Several further steps are required to transform silicon cell blanks into finished photovoltaic arrays, capable of absorbing sunlight and producing electricity.

---

*A recent review of technological status is given by McCleary [1].

**Cell thickness is assumed to be 0.15 mm and process efficiency following refining is assumed to be 40 percent.

At the earlier two of these steps (substrate preparation and junction formation) processes are fairly similar for silicon and gallium arsenide cells; the remaining stages (metallization, connection, and encapsulation) are common to all three photovoltaic technologies. We will discuss all these steps here as they apply to silicon cells; qualifications and exceptions pertaining to the other two technologies will be noted below, in the sections on cadmium sulfide and gallium arsenide. The particular secondary substances involved at these stages vary considerably, not only among the three generic technologies, but among variants of a single technology. However, a large number of the substances being used or considered are highly toxic; some are known or suspected carcinogens. The health effects of specific secondary substances (and combinations of substances) need to be studied in detail; their presence in the workplace and/or emissions to the environment should be carefully monitored and controlled.

After a silicon cell blank is cut from the purified silicon crystal or ingot, its surface must be carefully polished, cleaned, and etched. Next, the cell surface must be "doped"—that is, small amounts of selected impurities (such as boron, zinc, or selenium) are added to provide the excess or deficiency of electrons needed to make the cell a semiconductor. The substances used at this stage are typically highly toxic; they are typically applied as fine sprays or vapors.

The final stage in cell fabrication is the application of a fine metal conductive grid to allow the extraction of electrical energy in use. This is regarded by those engaged in photovoltaic research as potentially one of the most hazardous production steps, involving the presence of heavy metal or organometallic vapors and the use of other hazardous substances such as cyanides. Efforts are being made to develop metallization methods that allow isolation of workers at this stage.

Following metallization, the completed cells are assembled into panels or arrays, and their metallic grids soldered together. Finally, the cells are embedded in a pottant (usually a room-temperature vulcanizing material) and covered with glass or plastic and an antireflective coating. Potentially toxic or carcinogenic chemical agents, including organic solvents, are used in some of these steps, and heavy metal vapors or particulates from soldering operations may be present.

Too little is known about the precise character of processing risks at the various stages of cell fabrication and array assembly. These risks will depend on the detailed process chosen, the degree of automation, and the success achieved in implementing control technologies. However, considerable data could be collected from the existing electronics and semiconductor industries; it is important that such information be available on a timescale appropriate to the rate of development of a large-scale PV industry.

Cadmium Sulfide

Cadmium is currently largely a byproduct of zinc production,* with cadmium fumes released and partially recaptured when zinc and cadmium bearing ore

---

*Significant quantities of cadmium are also present in lead and copper ore but recovery is not common.

is roasted and sintered. Emissions from refining plants are significant, even in the U.S. Though uncertain, one estimate puts releases at about 300 pounds (see chapter 5) cadmium for each ton of cadmium throughput. About 70 percent by weight of this release involves particles of less than 2μ diameter, reflecting the low removal efficiency of control technology (baghouses and electrostatic precipitators) for removing very small particulates.* As discussed below, small particulates present special hazards. Cadmium recovered from ore roasting is reacted with hydrogen sulfide to produce a fine powder of cadmium sulfide (CdS); prior industrial experience (primarily in the paint industry) suggests that subsequent handling of this product also involves an appreciable fugitive dust problem.

Cadmium cell manufacture is more completely a series of chemical operations than the corresponding silicon cell process. Though there are several variations, the CdS cell consists of a thin film of CdS applied by vacuum deposition on a substrate, followed (or preceded, in some cases) by a dipped layer of cuprous chloride. Alternatively, the thin film of CdS may be produced by spraying a solution of $CdCl_2$ and thiourea ($NH-CS-NH_2$) on a heated substrate in air. The cells are completed by adding a very thin layer of copper and a transparent gold grid, followed by encapsulation. Workplace levels of CdS will probably be most significantly affected by the handling of input CdS powder; however, the fate of excess spray materials (cadmium and copper compounds) resulting from cell production should be examined. Problems arising in array fabrication will be similar to those encountered in silicon array manufacture; however, there is probably great opportunity for automation and closed process line operation for CdS cells, given current technologies.

The toxicity and carcinogenicity of cadmium and cadmium sulfide are still matters of some uncertainty. The industrial use of cadmium (electroplating, alloys, batteries, paint) has an unfortunate history, with chronic exposures of workers (largely through inhalation) and the public (largely through deposition on food crops) resulting in health damage that went unnoticed for decades. Cadmium metal and its oxide (fumes and dust) are now recognized as acutely and chronically toxic materials. Inhalation of fumes, as in refining or compounding, is most dangerous, followed by inhalation of dust and, finally, by ingestion. Acute doses (fatal pulmonary edema) have occurred at exposures of 2500 mg-min/m^3 (e.g., 8 mg/m^3 for five hours).[3]

Chronic exposures to levels lower than 100 μg/m^3 may result in emphysema, decreased bone mineralization, gastrointestinal disorders, and renal (kidney) damage, the latter leading to proteinuria and aminoaciduria. Increased cancer incidence (especially in the organs related to the renal system, such as the prostate) and hypertension have also been linked to low but extended exposure

---

*In principle, such controls can be very efficient, even for very small particulates; however, high efficiency requires high capital cost devices and careful maintenance. The latter can result in larger occupational exposures, illustrating an all-too-frequent tradeoff between occupational and public health impacts.

levels. The long residence time of cadmium in the body (a half-life of about 20 years) and the delay in health effects after exposure make difficult the determination of dose-response relationships.

Though cadmium and cadmium oxide are the dominant compounds encountered at the refining stage (and as a result of fires involving cadmium arrays), cadmium sulfide is the compound most likely encountered in the cell manufacturing industry.

Gallium Arsenide

Gallium arsenide cells are the least developed of the three cell technologies; process definition is at a very early stage with cell production still at the laboratory scale and based on proprietary methods. Thus, the following process description and discussion of potential sources of risk is general, reflecting the wide range of options and assumptions still associated with GaAs technology.

Gallium arsenide can be used in high efficiency single crystal cells consisting of relatively thick wafers (up to 300μ) since these cells operate well at high temperatures, over 200°F, they are best utilized in conjunction with Fresnel lenses or hyperbolic concentrators with concentration ratios ranging from 100:1 to 500:1. These configurations demonstrate an overall efficiency of up to 19 percent. The high power densities make it necessary to provide methods of dissipating excess heat from the concentrator-cell systems by using either forced air or water cooling, thus introducing the possibility of utilizing the waste heat. The high operating temperatures and cooling requirements suggest that these concentrator-GaAs cell configurations would be best suited for the larger commercial/industrial applications and for centralized power station systems.

Gallium arsenide can also be used in a polycrystalline form to produce lower efficiency (5-8 percent) thin film ($10-15\mu$) cells which may be incorporated in conventional flat plate assemblies suitable for residential and commercial/industrial use.

The direct impacts of producing gallium arsenide cells appear to be associated primarily with the large quantities of inorganic arsenic compounds which must be recovered, refined, compounded, and handled. Arsenic (as the trioxide) is obtained as a byproduct of primary copper smelting. Flue dusts from copper smelters are collected and As_2O_3 separated from other smelting byproducts. To obtain gallium arsenide of solar grade, arsenic trioxide must be reduced to metallic form which is then purified by distillation and compounded, at high temperature in a quartz crucible, with purified gallium, obtained from bauxite as a byproduct of aluminum refining. Between 25 and 200 kilograms of purified arsenic and a similar amount of gallium must be used to produce solar cells with a total peak output of one MWe. (The lower estimate is for single-crystal cells, the higher estimate for thin-film cells. See fig. 5.3.) A solar array manufacturing facility producing 200 MWe of peak capacity per year will use several thousand times as much gallium arsenide as the largest manufacturer of light-emitting diodes does now.*

Arsenic and its inorganic compounds are toxic and carcinogenic. Exposures at levels of 110 $\mu g/m^3$ have been reported to cause skin irritation, and higher levels lead to vomiting, nausea, diarrhea, inflammation and ulceration of mucous membranes and skin, and kidney damage. The effects of chronic arsenic poisoning include increased pigmentation of the skin, dermatitis, muscular paralysis, visual disturbances, fatigue, loss of appetite, and cramps. Liver damage and jaundice may result, as well as kidney degeneration, edema, bone marrow injury, and nervous system disorders. As more has been discovered about these effects, and as epidemiological data linking arsenic exposure to cancer incidence have been developed, regulatory standards for exposure have decreased precipitously. Occupational limits dropped from 500 $\mu g/m^3$ to 10 $\mu g/m^3$ in 1978 and further reductions—or, since practicality for existing industries was an issue in setting new standards, restrictions on new uses—are possible. There are as yet no public exposure standards. The implications of potential health impacts and regulatory processes for photovoltaic industries will be discussed in the sections on occupational and public health below.

Although the process steps used in the preparation of the GaAs wafer and subsequent cell fabrication are presumably similar to methods used in fabricating Si monocrystalline cells, there is not enough information about the proprietary processes involved to make a detailed health assessment. The production of polycrystalline GaAs cells poses problems and requires solutions similar to those associated with thin-film CdS cells. Array assembly involves the problems common to all three technologies, as discussed above.

Utilization

Photovoltaic systems may be used in a wide variety of applications and configurations, each of which may pose somewhat different risks to different groups of people. In terms of generating capacity, the applications of P.V. systems may be divided into three general categories:

- *Small scale 5-100 kw (peak);* decentralized, onsite application on residential structures ranging from single family dwellings to apartment complexes.
- *Intermediate scale 100 kw-1 Mw (peak);* decentralized, onsite service, commercial or industrial application (hospitals, colleges, shopping centers, office buildings, factories, government buildings, etc.)

---

*As of 1976, the largest LED manufacturer used 1 kg GaAs per month [4].

• *Large scale 10Mw-1000⁺MWe (peak);* central power applications ranging from community scale systems to large scale remote systems comparable to conventional coal and nuclear plants.

Small and intermediate scale systems may also vary in the ways in which they are integrated with other energy requirements and interfaced with existing utility services. For example, heat from photovoltaic systems may be collected and used for space or water heating, storage systems may be added to provide greater independence from existing grids, or utilities may be relied upon to provide extensive backup. The range of such possibilities is too great to be amenable to comprehensive and detailed analysis at this time. However, the precise ways in which photovoltaic systems are used will influence social (as well as economic) cost-benefit tradeoffs.

In the following chapters, we will focus on the problems which will be common to generic classes of photovoltaic technologies and applications as described above. In chapter 4 we review occupational health and safety issues; in chapter 5 we assess direct public health impacts; in chapter 6 we consider environmental impacts; in chapter 7 we evaluate the impacts of allocating energy, labor, and materials to the production and use of photovoltaic systems. Finally, in chapter 8 we summarize our analysis and explore its implications for the prospective development and use of photovoltaic systems.

REFERENCES

1. McCleary, J., *Photovoltaic Technology: A Review.* Cambridge, MA: MIT Energy Laboratory Working Paper No. MIT-EL 79-026WP, May 1979.
2. Since Si purification and crystal growth are very large cost elements, a variety of research is under way seeking means to modify or avoid these steps. Amorphous silicon looks increasingly promising for PV use. However, particular new processes must be carefully scrutinized for risk centers not present in conventional technology. For example, one experimental purification process, involving fluidized bed reactors, could cause worker exposure to large amounts of $SiCl_4$ and significant atmospheric release of particulates.
3. National Institute of Occupational Safety and Health, *Criteria for a Recommended Standard. . . Occupational Exposure to Cadmium.* Washington: U.S. Department of Health, Education and Welfare, Pub. No. (NIOSH) 76-192, August 1976.
4. *Inflationary Impact Statement, Inorganic Arsenic.* Prepared for the Occupational Safety and Health Administration by Arthur Young and Co. Washington: April 1976, p. A-32.

Chapter 4
Occupational Safety and Health Impacts

The use of a new energy technology—whether for conversion of coal to gaseous and liquid fuels or for the photovoltaic generation of electricy from sunlight—will entail a reallocation of societal resources and a reassignment of the costs of these resources. An important resource in all industrial activity is labor and among the costs of labor are the effects of that activity on the health and safety of workers. In evaluating photovoltaics on a social costs basis, it is necessary to consider how the manufacture and use of these technologies would affect the overall role of labor in society and also how the technologies would affect the spectrum of health and safety risks experienced by workers.

The first of these issues will be explored in chapter 7. Worker health and safety issues, dealt with in this chapter, include not only the social cost picture presented by emerging PV technologies, as compared to conventional options, but also the interaction of technological evolution with accumlating health effects data and changing regulatory standards. Since there are considerable differences in the policy significance of safety and health issues, we shall deal with them separately below.

In comparing energy technologies, two very different accounting perspectives may be used. The first is to compute the overall cost of a unit of energy produced by each technology in terms of worker injuries, illnesses, and deaths—much as one would do for public impacts. The second is to analyze the risk to individual workers active in each technology and to ask whether one technology presents greater individual hazards than another.

The first calculation gives the total cost to society—in terms of worker health and safety—of choosing a particular energy technology, while the second gives the cost to the worker of choosing a particular occupation. Both calculations are of value, but in different contexts. The societal cost calculation is useful in evaluating the benefits to society of particular energy choices and in considering possible unforeseen costs which may have to be borne by

15

society;* the individual worker risk is of interest to the worker in choosing a job and in evaluating prospective wage and working condition benefits.

It should be noted that these two calculations do not always give the same result: workers in a new energy industry may experience the same level of risk as those in a conventional industry, but the new technology may be considerably more—or less—intensive of labor than the old. Solar energy systems are generally more labor intensive than coal or nuclear power, per unit of energy output. Since the overall effect of this factor cannot be evaluated without also considering social costs and benefits other than occupational health and safety (such as reduced unemployment, reduction of dependence on imports, or alleviation of other risks to society), we postpone its discussion to chapter 7 of this study and focus here on the individual risk issue.

Occupational Safety

It is conventional to separate occupational safety from health considerations. Safety generally refers to accidents and deaths which are usually immediately evident and causally unambiguous. In contrast, health effects may be delayed in their appearance and only statistically correlatable (if that) with occupational exposures to toxic, carcinogenic, or mutagenic materials. Because safety hazards are relatively easily identified and responsibility for them assigned, there are economic and other pressures which lead to their alleviation or internalization. To the extent that risks cannot be removed or internalized as economic factors directly by individual firms or industrial sectors, there are workmen's compensation and other insurance programs.

The fact that safety can usually be translated directly into economic terms appears to be responsible for the lack of significant variation in safety hazards between industries. For example, injury rates for various manufacturing industries range from about five injuries per 100 worker-years (as for semiconductor manufacturing) to about 30 (for non-ferrous metals production).[1] Accidental death rates in construction range from about two to eight per 10,000 worker-years.[2] One should thus expect that the safety hazards associated with manufacturing and installation in a mature photovoltaics industry will fall within the same range as other industries and energy technologies, and that the corresponding costs will be internalized.

Within this range, the specific characteristics of risks in a future PV industry may be estimated by comparison with similar activities already occurring. Thus, silica mining may be compared with general mineral mining statistics (6.8 injuries per 100 worker-years); refining and processing of cadmium and arsenic may be compared with non-ferrous metals industries (29.2); cell manufacturing

---

*An example is the national programs of support for those affected by coal mine pneumoconiosis.

and array assembly with chemical process industries (8.5), semiconductor industries (6.0), and electrical equipment manufacturing (6 to 17); and installation of decentralized arrays with electrical work (15.8) and roofing and sheet metal work (25.6).* In comparison, petroleum and coal process industries involve 6 to 17 injuries per 100 worker-years. These statistics do not entirely reflect the spectrum of severity of injuries involved—for example, injuries for coal miners, nuclear plant construction workers, or roof workers are probably more serious than those in the semiconductor industry—but do suggest, under the reasonable assumption that photovoltaic industrial activity will not present individual safety risks very different from those in comparable trades, that the average worker risk for the photovoltaic industry as a whole is likely to be similar to the risks involved in conventional energy systems. Indeed, it is possible that avoidance of heavy construction activities for conventional technologies (cooling towers or containment domes), which have injury and death rates near the upper end of industrial risk scales, would result in a lower average individual worker risk for photovoltaic systems.

Health Impacts

While safety problems are readily detectable and attributable—and thus internalized in economic and other decision processes—the same is not true of health effects. Occupational health effects may result from short-term or chronic exposure to toxic, carcinogenic, or mutagenic materials. Whereas safety hazards generally lead to immediately evident injuries or deaths, health hazards may be characterized by effects that are slow to develop, cumulative, and complicated by non-occupational factors. The latter may include synergistic effects, arising from the interaction of workplace chemical exposures and general environmental pollutants or other factors to produce disease which would not occur in response to either factor alone.** Very little is known about the number and magnitude of synergistic health effects; they are frequently characterized by delays in the manifestation of symptoms.

The health consequences of worker exposures are thus difficult to foresee, detect, or quantify and even more difficult to attribute causally to the workplace.*** The result, in the past, was that health effects were only rarely internalized; costs were instead involuntarily (and generally unknowingly) borne by individual workers, or shared, to some extent, through health insurance

---

*The high rates for roofing and sheet metal work appear to reflect much higher rates for hot composition roofing and a large proportion of novice workers.

**A relevant example is the increase in ill effects of cadmium in individuals suffering from calcium deficiency.

***It is significant that only a few percent of reported injuries and illnesses nationally are in the illness category and of these about 70 percent are readily diagnosed skin ailments [1].

or social welfare systems. Internalization of prospective occupational health costs in economic decisions—say, in deciding to install costly workplace controls—would require an accurate system for bookkeeping exposures over a worker's career with multiple employers, a method of assigning responsibility for effects which might have more than one cause and for anticipating—at least on average—the extent of health impacts which are actually unknown or have substantial latency periods.

Because of the complexity of such calculations—and perhaps even more because of the public's perception that employers may not voluntarily go to any great lengths to protect workers—government has in recent years become increasingly involved in monitoring and regulating occupational health hazards. Where relevant, regulatory issues are discussed in the subsections below which deal with the particular occupational hazards posed by photovoltaic technologies. But because the current regulatory climate is likely to affect profoundly the development of any new technology, we discuss more generally in the concluding section of this chapter the regulatory philosophy now prevailing and its implications for photovoltaics.

The following subsections explore the specific potential hazards associated with manufacturing silicon, cadmium sulfide, and gallium arsenide photovoltaic systems. Because the detailed processes of large-scale production are either not yet defined, or are proprietary, the primary focus is on the flows of basic chemical consituents of PV systems: silicon, cadmium, and arsenic. With the possible exception of conventional structural materials (glass, aluminum, and so forth), these elements are also likely to be the source of most hazards affecting appreciable numbers of workers, due to the large throughputs and the high-temperature processes involved.* It should be noted, however, that our analysis is not a substitute for a program of continuing examination of occupational hazards in an evolving industry and that technological change may reshape the structure of risks identified here. Furthermore, as research continues on photovoltaic suboptions, it must be understood that no particular production process can be deemed commercially and socially viable without detailed consideration of (1) the specific occupational health risks introduced at every production step; (2) the permitted levels of exposure for all materials and combinations of materials involved, and the likelihood of change in relevant regulations; and (3) the feasibility and costs (economic and other) of measures to control or reduce the occupational health hazards of that specific solar technology.

Silicon

The primary locus of occupational health issues for silicon photovoltaic cells is silica dust and other silicon compounds. Of greatest concern is silicon refining

---

*Social costs and benefits associated with use of conventional structural materials in photovoltaic devices are discussed in chapter 7.

with electric-arc furnaces; other potential exposures occur in mining and through use of trichlorosilane to purify metallurgical-grade silicon further for solar use. Silica was one of five substances initially chosen by OSHA for special attention because of its demonstrated role in health problems and because of the large number of workers affected (about 1 million)[3]. In the form of fine crystalline particulates, silica is highly dangerous; once inhaled it can result in scarring of lung tissue (silicosis) or be translocated to the kidneys or other sensitive organs. Synergistic effects involving other pollutants are also likely (e.g., particulates may provide catalytic surfaces for reactions involving chemical pollutants or a transport mechanism into the lung) and impairment of lung function may increase the likelihood of other diseases.

Whereas the OSHA standard for the respirable fraction (less than 7 microns) of unspecified particulates is set at $5mg/m^3$, standards for various forms of silica containing particulates are set considerably lower, as shown in table 4.1. The most restrictive standard is for crystobalite—a particular crystalline form likely to be produced at high temperatures. Present standards do not distinguish submicron from larger particulates, though it is likely that more restrictive standards for the submicron component will eventually be promulgated.

It is generally believed that the mining of silica—the basic raw material for silicon cells—is not a source of particulate emissions of significance to occupational (or public) health. This belief appears (e.g., see Lockheed[4]) to be based on a study by Cholekoda and Blackwood[5] and on extrapolation from the sand and gravel industry to estimate the number of workers at risk in mining silica. In fact, mining of the high-grade silica (98 percent SiO_2) required for silicon production is about 60 times as labor intensive as the sand and gravel industry.* Production of 10 GWe-peak of silicon photovoltaic cells, using current technology, would require about twice as much high-grade silicon as is currently produced and would greatly increase the number of workers involved. Furthermore, while there appears to be no evidence that mining of high-grade silica presents inordinate individual worker hazards, the study usually cited does not provide proof that such hazards are absent; its methodology is questionable in several important respects.[6] These problems leave open the question of actual worker health hazards in silica mining—a question which deserves more careful review.

The reduction of silica to metallurgical grade (96-98 percent pure) silicon is a significant source of SiO, SiO_2 particulate and fly ash emissions to the environment. Between 20 percent and 40 percent of workers involved in the production of silicon photovoltaic arrays are employed at this stage. A large proportion of the emissions from the reduction of silicon alloys are submicron particulates; for example, EPA measurements of *mean* particle diameter ranged

---

*Total sand and gravel production in 1973 was 938×10^6 tons, while 0.589×10^6 tons of high-grade silica were mined; about 50,000 workers were employed in the domestic sand and gravel industry, while 2000 were employed in mining high-grade silica.

Table 4.1. Federal Standards for Particulates Containing Silica*

Species	Maximum Concentration (mg/m^3)
Amorphous Silica	$\dfrac{80}{\text{Fraction } SiO_2 + 2}$
Total Crystalline Silica	$\dfrac{30}{\text{Fraction } SiO_2 + 2}$
Respirable Crystalline Silica	$\dfrac{10}{\text{Fraction } SiO_2 + 2}$
Crystobalite	$\dfrac{5}{\text{Fraction } SiO_2 + 2}$

*Occupational Safety and Health Standards, (Sec. 1910.1000 − (d) (1) (i), April 20, 1978, p. 42).

from 0.66 to 1.7μ for the emissions from the open furnace production of ferrochrome silicon.[7] As the silicon content of the final product increases, the rate of particle formation increases and the particle size decreases. The result is that particulate emissions from the production of silicon metal are the hardest to control.

The silica emissions from the arc furnace pose a special danger for two reasons. The high temperature process tends to produce SiO, a highly reactive compound whose presence may imply a greater hazard to workers in the immediate vicinity of the furnace than would SiO_2. Second, the silica particulates emitted from the arc furnace are likely to be tridymite and crystobalite which are chemically identical to quartz but differ in crystalline structure. More hazardous than all other forms of silica, crystobalite and tridymite are formed at the high temperatures present in the arc furnace.

Open and semi-closed electric arc furnaces allow emissions of such compounds directly into the workplace—a problem that could not occur with totally enclosed arc furnaces. However, at this time silicon metal is produced only in open arc furnaces because of the need for periodic breaking of the crust covering the charge, a necessary step in order to prevent violent gas jetting and bridging of the electrodes. Large amounts of ventilation are required to keep temperature and particle concentrations at comfortable and safe levels. While high air flow rates reduce worker exposures, they may make it more difficult to control emissions to the atmosphere, as discussed in the public hazards section.

The extent of occupational hazard associated with particulate emissions from the arc furnace reduction of silicon will depend on resolution of several key areas of concern:

- *Uncertainty about the nature of the emissions*—the amount of SiO which may reach a worker is unknown; although SiO oxidizes to SiO_2 in air, direct emissions from open arc furnaces may expose workers to significant amounts of SiO; particle size and characteristics (whether amorphous or crystalline) must be better known.

- *Appropriateness of silica standard*—SiO is potentially more hazardous than SiO_2 but this is not reflected in the silica standards; biological effects of crystalline and especially of sub-micron particulates are still poorly understood; silica standards do not take special account of sub-micron particulates.

- *Effectiveness of the control technology*—present control technology is very energy intensive (large amounts of air must be moved); particulates have high electrical resistivity and therefore are not easily controlled by electrostatic precipitators; high air flow rates make atmospheric emissions control more difficult; discontinuous emissons due to periodic stirring and tapping pose special difficulties.

- *Feasibility of closed electric arc silicon furnaces*—a closed furnace has potential for reduced particulate emissions and reduced heat loss; it requires less than 5 percent of the air flow rates needed in open arc furnaces; therefore, it also has significantly lower energy needs.

The arc furnace reduction process also requires significant amounts of coke—about 21 kg for each kilogram of cell-grade silicon ultimately produced. Although the old bee-hives once used to produce coke furnaces are almost completely replaced by new units with much lower atmospheric particulate emissions, coke production is still a focus for concern, with OSHA trying to reduce the carcinogenic benzene soluble fraction of total particulate matter to 0.15 mg/m^3 (eight-hour average in the workplace). Additional data and analysis will be required to establish the health impacts attributable to silicon photovoltaics due to these releases in coke production.[8]

After metallurgical grade silicon is produced, its purification to semiconductor-grade Si metal requires the processing and handling of large quantities of halogen compounds such as $SiCl_4$ and $SiHCl_3$. Although not particularly toxic, the physical and chemical properties of these compounds make them quite sensitive to changes in temperature and reactive when exposed to even small amounts of air or moisture. Thus, these compounds could present hazards in case of accidental spills and leaks, the result of which could be rapid decomposition and generation of large quantities of toxic chlorine gas. However, these chemicals have been utilized in large quantities for a long time in other industries with few significant problems. The only other potential problem identified at the refining stage is chronic exposure to concentrations of HCl emitted during the purification process; the OSHA standard for HCl is 7 mg/m^3. Beyond the refining stage, the only potential worker exposure to significant levels of silicon particulates is during the cropping and sawing of wafers. However, these processes are likely to be enclosed and automated, greatly reducing or eliminating human involvement.

Cell Fabrication and Array Assembly

Fabrication of cells is another area where workers may be exposed to significant health hazards. As noted in chapter 3, the three photovoltaic technologies involve rather similar problems at this stage, and resemble each other still more closely at the array assembly stage. As we did in chapter 3, we will discuss these production steps here as they apply to silicon cells; variations pertaining to the other two technologies will be noted below.

Although extremely small amounts of toxic phosphine (PH_3) and boron trichloride (BCl_3) dopants are used to form semiconductive junctions for silicon cells, care should be taken to determine and achieve safe exposure levels for nearby workers. Current OSHA standards for phosphine are 0.4 mg/m^3 (eight-hour time weighted average); chronic exposure to low levels of phosphine is reported to lead to anemia and various nervous system disorders.[9] The lethal concentration for BCl_3 is 20 ppm. Etching of the silicon wafer is done with HF which results in the production of a highly toxic gas, SiF_4. Projections of actual hazard levels can only be made when engineering design studies are available.

Once the cells (whether based upon silicon, cadmium sulfide, or gallium arsenide) are made semiconductive, metallic grids must be applied and cells must then be assembled into arrays. At present, metallization techniques allowing complete worker isolation are still being sought, and array assembly involves extensive fabrication by hand; at both these steps there is possible worker exposure to toxic or carcinogenic chemical agents, including organic solvents and heavy metal vapors (or small particulates) from soldering. Similar conditions exist in the industries involved in the manufacturing of industrial and scientific instruments, fabricated metal products, and the manufacturing of electrical equipment and supplies. These industries are rated by the National Institute for Occupational Safety and Health as the three most hazardous industries in the United States, in terms of exposure of workers to carcinogens.[10] Since OSHA recommends complete isolation of workers from carcinogenic substances, present efforts to automate the metallization and fabrication of photovoltaic cells and arrays are highly desirable from the standpoint of occupational health.

Cadmium

Cadmium can be refined from the collected particulate releases from copper and lead as well as from zinc smelting, but most cadmium is recovered from the latter. At the smelter, flue dusts containing small particulates of cadmium oxide are collected from baghouses or electrostatic precipitators. The collected dust is then purified by a sequence of chemical processes ending with the production of cadmium metal sponge through a distillation process. Throughout this sequence, there is potential for occupational exposures to fine dusts of cadmium and cadmium fumes (wherever high temperatures are involved). Following purification, cadmium has been used primarily for electroplating or batteries or compounded with other elements (notably sulfur) to make pigments. Health experience in these industries is reviewed in a recent NIOSH criteria document.[11]

A new industry for cadmium sulfide cell manufacture may pose problems similar to those encountered in traditional industries, though assessment is made difficult by the proprietary nature of present processes. Control problems will arise at two steps (at least): the initial handling of cadmium sulfide (a fugitive dust problem) and the control of vapor used in vapor deposition. The precise character of these and other problems should be examined as part of the general development program for the technology.

The evolution of a cadmium photovoltaics industry would very likely be shaped by growing knowledge of the health effects of cadmium. Studies of workers and others exposed to cadmium, as well as a great many animal experiments, have revealed a large number of potential health impacts. Some of these—such as proteinuria due to effects on the kidney and respiratory disease—are well-demonstrated. Others—such as hypertension, carcinogenic, and mutagenic effects—have not been proven, though there is suggestive evidence. For example, a number of animal experiments, and a few epidemiological studies have correlated the presence of cadmium compounds with hypertension and other circulatory problems; a study by Lemen[12] of workers in a facility producing very pure cadmium oxide and cadmium sulfide indicated excesses of total malignant neoplasms (respiratory and prostate tumors), despite use of respirators; and cadmium sulfide has been shown to induce chromosomal aberrations in cultured human leukocytes.[13] While the data cannot yet be regarded as conclusive, it is at least possible, if not probable, that future health assessments—and regulatory standards—will have to consider effects which can be correlated statistically with even low dose levels.

As of April, 1978, the Occupational Safety and Health Standards on Toxic and Hazardous Substances list eight-hour time-weighted average exposure limits of 200 μg/m^3 for cadmium dust and 100 μg/m^3 for cadmium fumes.[14] While industry appears to have been able to meet these standards, chiefly through improved ventilation, there is an increasing likelihood that the standards will be made more restrictive. For example, the NIOSH Criteria document recommends that worker exposures should remain below 40 μg/m^3 (TWA) and that no 15-minute exposure should exceed 200 μg/m^3. Even more stringent standards are possible, especially if evidence for carcinogenicity is sustained,* since the trend in regulation is to regard *any* exposure to carcinogens as

---

*The NIOSH Toxic Substances List of 1976 includes cadmium and cadmium sulfide as carcinogens, based on animal studies.

unsafe. The prospect, in this case, is for increasingly severe ratcheting on cadmium standards.*

If cadmium is to become a commercially viable photovoltaic technology, it will be necessary to anticipate prospective regulatory changes through evaluation of health effects and through industrial and process designs which provide very high levels of control. The latter will be difficult in the case of the high temperature processes used at several points in the chain leading from ore to high purity cadmium sulfide. At the cell manufacture and array fabrication steps, the need for control would reinforce economic incentives to establish closed automated process lines. The more difficult questions are whether control at levels which deal adequately with health impacts and future standards is possible and whether industrial and regulatory management will in fact implement appropriate measures.

A final issue which must be resolved is whether cadmium sulfide presents health risks comparable to or less than those associated with other cadmium compounds. While cadmium and cadmium oxide are the compounds encountered at many early stages of processing (and as a result of accidental releases, where heating in air may result in CdO formation), cadmium sulfide is the compound most likely encountered at the cell and array fabrication stages. It is frequently asserted that cadmium sulfide is ten times less toxic than cadmium. We have been unable to find adequate documentation for this assertion, despite a computer search of the NASIC Toxline reference files. While there is evidence that the lethal dose of cadmium sulfide is greater than that of the oxide,[15] the data are for ingestion rather than inhalation; the latter is the only hazard of realistic concern in photovoltaic manufacturing. Other studies which have been used to argue for lower risk from CdS deal with inequivalent systems; for example, tumorogenesis following intramuscular injection in rats as compared with toxic oral doses. The effects of *inhalation* of cadmium sulfide by humans should be reviewed much more exhaustively** before it can be concluded that standards for cadmium sulfide can be set higher than for other cadmium compounds. Until further research proves otherwise, it is sensible to follow the NIOSH recommendations*** and treat cadmium and all cadmium compounds equally.

---

*The use of cadmium photovoltaic cells on the scale anticipated by DOE goals would increase domestic use of cadmium by about a factor of four over present levels by the year 2000, though this figure depends critically on cell characteristics and process efficiencies.

**There is some evidence, from an old study, that cadmium sulfide is absorbed somewhat less readily in canine lungs than cadmium oxide.[16]

***The NIOSH Criteria document refers to "elemental cadmium and all cadmium compounds" and finds no basis in health data for distinction.

Gallium Arsenide

Of the photovoltaic technologies discussed here, gallium arsenide is the least developed, just emerging from the laboratory to be considered for large-scale production. As a result, it is difficult to assess any hazards except those associated with the key material constituents, gallium and arsenic.

Very little is known about the health-related properties of gallium, a by-product of aluminum refining. According to the NIOSH Registry,[17] the lethal dose for gallium is 110 mg/kg (subcutaneous in rats); it is not listed there as a carcinogen, though some reports [4, p. 106] refer to it as such. Gallium is used as a scanning agent in medicine and has been proposed as a malleable constituent for dental amalgams (for which mercury was previously used). A review of Toxline references reveals virtually no reports on health effects, though this should certainly not be taken as evidence of safety.

There is considerably more data on the health effects of arsenic and arsenic compounds, data which have led to increasingly strict occupational standards. In addition to known toxic effects, arsenic is now regarded as a carcinogen, based largely on epidemiologic studies. (Unlike the case with most carcinogens, animal models have proven more elusive than data on human populations, perhaps because of metabolic differences.)

Until recently, the occupational limit (eight-hour time-weighted average) for exposure to arsenic, in organic and inorganic compounds, was 500 $\mu g/m^3$. A NIOSH Criteria report, in 1974, recommended a reduction for inorganic compounds to 4 $\mu g/m^3$, based on a comprehensive review of health data. In May, 1978, OSHA promulgated a new standard for exposure to inorganic arsenic (organic compounds, used in agriculture, remain at the higher standard) of 10 $\mu g/m^3$ (eight-hour time-weighted average). The text of the regulation states that:

The basis for this action is evidence that exposure to inorganic arsenic poses a cancer risk to workers. The purpose of this rule is to minimize the incidence of lung cancer among workers exposed to inorganic arsenic. Employees protected by this standard work principally in the nonferrous metal smelting, glass and arsenical chemical industries. Provisions for monitoring of exposures, recordkeeping, medical surveillance, hygiene facilities and other requirements are also included. The 10 $\mu g/m^3$ limit has been set because it will provide significant employee protection and is the lowest feasible level in many circumstances.[18]

Thus, the OSHA regulation represents a compromise between feasibility (the 10 $\mu g/m^3$ average limit will probably still require severe control measures and rotation of workers) and its philosophy that there is no zero risk level for carcinogens. However, the new standard probably does reduce the individual worker risk and, to a more limited extent, the risk to arsenic workers collectively.

As is the case with cadmium, worker exposure to arsenic is possible both during the production of the raw material and during the subsequent manufacturing steps. The history of worker exposure during the recovery of arsenic from refinery flue dusts in the U.S. has been unfortunate; compliance with the new standard should reduce this problem considerably. Any expanded production due to photovoltaics should be planned with careful consideration given to minimizing worker exposure.

Most gallium arsenide now produced is used in light-emitting diodes. A survey of this semiconductor industry in 1976 indicated average workplace exposure levels of 5 to 50 $\mu g/m^3$ of arsenic, as estimated by company spokespersons.[19] Concentrations in crystal growth areas are probably considerably higher, due to daily clean-up operations [19, p. IV-19]. It is likely that ventilation will keep levels in the diffusion process area (where commercial arsenic trioxide is purified to semiconductor grade) below the new standard; air-supplied respirators may be necessary in crystal growth areas.

The extent to which experience with current semiconductor industries can be generalized to large-scale photovoltaic operations is limited. A facility producing 200 MWe-peak of gallium arsenide photovoltaic systems would use about 10,000 times as much gallium arsenide as is now used by the largest GaAs semiconductor manufacturer. Control problems in the photovoltaic industry may thus be of an entirely different nature and scale than those encountered in present industries. However, it does seem clear that occupational exposure problems may be more easily solved with thin film gallium arsenide technologies which can be adapted to closed process line operations than with crystal wafer technologies which require closer human involvement.[20] The production process for polycrystalline GaAs cells will have to include methods to deal with the existence of excess spray materials from the vacuum deposition and spray application of thin films.

The fabrication of GaAs arrays presents the hazards that are common to all the PV technologies (presence of cutting oils, solder compounds, materials used in vapor deposition of contact grids) as well as unknown hazards related to the proprietary compounds used during GaAs compounding, doping, and crystal growth. Again, the unknown synergistic behavior of these various compounds makes it highly desirable to insure isolation of the various production phases from the workers' environment.

Installations

The use of photovoltaic systems on buildings or in independent arrays may alter the nature and extent of health hazards to occupational groups not directly involved in the installation and maintenance of photovoltaic systems. Such effects are probably most significant for firefighters and other emergency personnel. Not only may such workers be repeatedly exposed to toxic materials or electrical hazards, but the presence of photovoltaic systems on roofs may affect the ways in which fires can be fought.

Toxic material releases will be most significant for gallium arsenide since the volatility of arsenic compounds insures high release fractions at temperatures which are low compared to those commonly achieved in building fires. Due to sublimation and oxidation processes, it is likely that primary exposures would be to vapors or small particulates of arsenic trioxide. Release fractions for cadmium sulfide arrays will be considerably lower than for arsenic based cells, due to a higher temperature for sublimation. Experimental studies will be necessary to determine whether arsenic or cadmium uptake from a single fire could lead to prompt toxic or even lethal exposures for firemen. The possibility of exposures to toxic materials resulting from combustion of other materials, such as pottants, used in PV systems, also deserves study. In this connection it should be noted that toxic gases are already a serious hazard; a recent study conducted jointly by the Harvard Medical School and the Boston Fire Department recommends the use of air-supplied respirators for firemen. Such measures would reduce cadmium and arsenic hazards considerably.

The possibility of cumulative exposures to firemen and other emergency personnel is more difficult to assess since ambient concentrations near fires may not be great enough to justify respirators but high enough to lead to long-term accumulations. A rough calculation can suggest a measure of the hazard. Assuming an average exposure of about 0.01 g-min/m^3 of cadmium or arsenic per fire (see chapter 5) and 300 such exposures in a career, a firefighter would retain a total of less than 0.02 gram of arsenic or cadmium. For cadmium, this is less than one-fourth the quantity needed to cause kidney function impairment. While exposure to cadmium or arsenic may increase individual cancer risk slightly, exposures would be less than those presently allowed for workers in industries using either element. However, the calculations here suggest that the question of long-term effects on emergency personnel deserve further attention. Silicon photovoltaic systems do not raise these questions.

The presence of photovoltaic systems on roofs may also alter the nature of hazards faced by firemen by requiring changes in the ways in which fires are fought. As discussed in chapter 5, interior fires are often shaped into a vertical configuration by opening holes in the roof. This minimizes the lateral spread of the fire, reduces hazards to entering firemen or rescue personnel, and minimizes property damage; it also makes roof involvement much more likely. If the value of photovoltaic systems, or the risk of releasing toxic materials, inhibit use of current firefighting techniques, it is possible that conventional hazards to firefighters will be increased.

Occupational Health Standards
and Regulatory Changes

The course of development of a technology is determined not only by the risks known (or suspected) to be associated with it, but also by the attitudes held by society toward various kinds of risks and by the institutions created to implement those attitudes. In the United States, the 1970 Congressional Act establishing the Occupational Safety and Health Administration was a major extension of government's role in protecting the welfare of workers. With respect to allowable concentrations of toxic, carcinogenic, or other hazardous substances in the workplace, the Act stipulates that OSHA must set the standard "which most adequately assures, to the extent feasible, on the basis of the best available evidence, that *no employee* will suffer material impairment of health or functional capacity even if such employee has regular exposure to the hazard for the period of his working life" [18, p. 19585].

The 1970 Act is considerably more restrictive than earlier standards: its aim is to see that no appreciable harm comes to any worker, whereas the (rather imprecise) Threshold Limit Values on which previous standards were based aimed at ensuring less than one death per million "normal healthy workers." OSHA's task in establishing actual standards is immense: 12,000 toxic chemicals are currently in use, and 500 are added annually. Adequate human dose-response data are currently available for very few of these substances, and for almost none of the millions of combinations which could interact to cause harmful effects.*

Perhaps the most far-reaching aspect of the Occupational Safety and Health Act is its requirement that standards be based on "the best available evidence." In the past, TLV's were often based on extrapolation from animal experiments; now the trend is to rely on direct human data, on specific epidemiological studies, and on toxicological analyses to discover the chemical interactions between toxic substances and human metabolism. It seems clear that the intent of the Act is that permissible exposure levels should be changed whenever there is reliable new evidence about concentrations at which a given substance is hazardous to humans. The Act's fundamental conservatism on

---

*Carcinogens pose special problems. As longer periods of exposure are investigated (30–50 years), more and more toxic substances are found to correlate with increased cancer incidence at low levels of concentration. OSHA's present policy is to assign absolute levels of exposure for carcinogens because they are required by law to do so. However, OSHA and NIOSH have publicly expressed the belief that any level of exposure to carcinogenic agents—including levels not measurable by present technology—can be responsible for significant health effects. In some recent comments about rule-making on carcinogens, NIOSH stated that "health risk should be given primary consideration in questions of feasibility. . . ." Furthermore "if the lowest feasible limit cannot be set so that it is lower than that concentration which has been found to cause cancer in humans and/or animals, then exposure in the workplace should not be permitted." [21]

behalf of worker health makes it overwhelmingly likely that most if not all changes will be in the direction of tighter restriction.

The implications, for both new and existing technologies, are clear. Any technology, no matter how benign at first glance, may lead to the introduction of significant concentrations of substances with the potential for adverse occupational health impacts. Establishing standards for allowable exposure to harmful substances requires considering numerous factors: biological susceptibility, synergistic effects, carcinogenic behavior—all presently associated with high uncertainty. As more information becomes available about the long-term effects of toxic and carcinogenic substances, and as efforts are made to deal with the wide range of susceptibilities in the worker population, it is very likely that standards for occupational exposure will become increasingly strict and that the overall trend will be toward more complete isolation of the worker from materials which may pose health hazards. Therefore, in any industry, present standards must not be seen as absolute; they should be viewed with the understanding that their inherent uncertainty requires use of an additional margin of safety and that this apparent margin can be eroded by better health data.

This current regulatory climate is of great importance in assessing the prospects for photovoltaics and in choosing between technological suboptions. A new industry is especially vulnerable to regulatory changes since it may require greatly expanded use of materials about which there is already escalating concern, and hence a great reluctance to allow wider use, or because the new industry may lack the political support which, in established industries, can moderate the imposition of rapid changes in standards, especially in cases where scientific evidence is not entirely conclusive. Indeed, the increased concern over possible effects on health may result in the subjection of PV technologies to a level of scrutiny not imposed on technologies introduced in an earlier health assessment era, although it is to be hoped that present energy technologies will not be exempted from the same careful analysis.

Our analysis above suggests that silicon-based photovoltaics involves fewer (and more tractable) occupational health risks, and thus is less vulnerable to regulatory intervention and change, than is the case with technologies based on cadmium or gallium arsenide. However, worker health hazards exist for all three technologies; these hazards must not only be assessed in detail, they must also be eliminated when possible and in other cases must be understood and accepted by the public and by the workers involved before large-scale commitments are made to particular production sequences.

REFERENCES

1. "Occupational Injuries and Illnesses in the United States by Industry," *Bulletin 1932*. Washington: Bureau of Labor Statistics, U.S. Department of Labor, 1974.

2. *Accident Facts.* Chicago: National Safety Council, 1972.
3. *Criteria for a Recommended Standard . . . Occupational Exposure to Crystalline Silica.* National Institute of Occupational Safety and Health. Washington: U.S. Department of Health, Education and Welfare, 1974.
4. Gandel, M.G., et al., *Assessment of Large-Scale Photovoltaic Materials Production.* Lockheed Missiles and Space Company, Inc. Prepared for the Environmental Protection Agency, EPA-68-02-1331. Cincinnati: August 1977.
5. Chalekode, P.K. and Blackwood, T.R., "A Study of Emissions from Aggregate Processing." Presented at the 69th Annual Meeting of the Air Pollution Control Association, Portland, Ore., June 1976.
6. Emission factors were derived by measuring ambient concentrations some distance from sources and extrapolating back to actual working areas using a dispersion model—concentrations in work areas were not actually measured. The silica content of dusts to which workers were exposed also was not determined directly, but rather inferred from measurements made on samples collected from unpaved quartzite mine roads. Finally the silica content (17 percent) was determined for the *respirable* fraction of road dust but the permissible exposure used as a reference health standard was for *total* silica particulates—a factor of three too high.
7. *Production of Ferroalloys,* Vol. 1. Environmental Protection Agency, 450/2-74-018a. Washington: 1974.
8. *Occupational Health and Safety.* Notice regarding coke emissions standards. January/February 1977, p. 9.
9. *Solar Program Assessment: Environmental Factors, Photovoltaics.* Environmental and Resource Assessments Branch, Division of Solar Energy, ERDA 77-47/3/ Washington: Energy Research and Development Association, March 1977.
10. Maugh, T.H., "Carcinogens in the Workplace: Where to Start Cleaning Up," *Science,* Vol. 197, 1977, p. 1268.
11. *Criteria for a Recommended Standard . . . Occupational Exposure to Crystalline Silica.* National Institute of Occupational Safety and Health. Washington: U.S. Department of Health, Education and Welfare, Pub. No. (NIOSH) 76-192, August 1976.
12. Lemen, R.A., et al., *Cancer Mortality Among Cadmium Production Workers.* Ann NY Acad Sci 271: 273-79, 1976. As cited in *Occupational Exposure to Cadmium,* NIOSH, August 1976, p. 17.
13. Shirashi, et al., *Proceedings of the Japanese Academy.* 48:133–137, 1972.
14. "Occupational Safety and Health Standards," *OSHA Reporter Reference Manual.* [Sec. 1910.1000 (d) (1) (i)].
15. Vorob'eva, et al., *Gig. Sandit.,* Iss. 2, Moscow, 1975, pp. 102–104.
16. Princi, F. and Geever, E.F., "Prolonged Inhalation of Cadmium," *Archives of Industrial Hygiene and Occupational Medicine 1.* Chicago: American Medical Association, 1950.
17. *Registry of Toxic Effects of Chemical Substances.* National Institute of Occupational Safety and Health. Washington: 1976, p. 559.

18. "Occupational Exposure to Inorganic Arsenic—Final Standard," *Federal Register,* Vol. 43, No. 88, May 5, 1978.
19. *Inflationary Impact Statement, Inorganic Arsenic.* Performed by Arthur Young and Co. Washington: Department of Labor, Occupational Safety and Health Administration, April 1976.
20. There may also be accidental release hazards which should be considered. For example in the manufacture of thick wafer monocrystalline cells, the GaAs ribbon pulling process is conducted at high temperatures (1250° C) and pressures (100 atm) and therefore introduces the possibility of chronic low level leaks or accidental exposure to high arsenic emissions from the reactor vessel. The alternate epitaxial growth process may be less prone to accidental releases.
21. *OSHA Reporter,* Vol. 7, No. 43, March 23, 1978, p. 1601.

Chapter 5
Public Health Impacts

Public health risks arising from photovoltaics may be direct or indirect. Direct impacts include routine or accidental releases of toxic or carcinogenic materials during manufacturing and utilization of photovoltaic systems. Indirect impacts include the effects of producing the inputs to photovoltaic manufacturing processes: aluminum, solder, plastics, glass, and so forth, and the energy it takes to produce the photovoltaic system itself.

In the following subsections we shall examine the potential direct public health impacts of the three main lines of photovoltaic technologies. Discussion of how indirect impacts affect the comparison between energy technologies is reserved for chapter 7.

Nature and Sources of Direct Impacts

As discussed in chapter 3, photovoltaic systems involve the presence of potentially hazardous materials at several stages of production or utilization. Atmospheric releases of these materials appear to be the primary source of public exposures. The dominant effects in silicon technology appear to come from the mining and refining of silicon, including the use of coke (as a reducing agent) and the release of silicon particulates. For cadmium technology, the major public risks appear to come from the release of cadmium compounds in refining, during processing and handling in manufacture, and from fires involving arrays. Risks associated with gallium arsenide are qualitatively similar in their origin to those of cadmium photovoltaic systems. We shall examine each of these effects in turn.* In order to maintain perspective, however, it is useful to consider how comparison can be made with conventional sources of

---

*We shall make a distinction here between immediate public exposures due to photovoltaics and those health effects which may eventually result from contributions of pollutants to environmental backgrounds. The latter effects will be treated in the subsequent chapter on environmental impacts.

32

electricity. The comparison with nuclear power is made difficult by the incomparability of risks (low-probability high-consequence events, nuclear weapons proliferation, and so forth). However, it is possible to make more direct comparison with the effects of using coal.

In addition to the well-known releases of oxides of sulfur, nitrogen, and carbon, the mining and combustion of coal also involve the emission of particulates (including silicon), cadmium and arsenic—precisely the materials emitted in the use of photovoltaics. Though there are some differences in particle size distribution and chemical and molecular composition, it should thus be possible to evaluate public health impacts of photovoltaics on a basis which allows reasonably direct comparison with a subset of the impacts of coal. For example, if it can be shown that the human exposures to silicon, cadmium, or arsenic resulting from the use of photovoltaic systems are smaller than or comparable to those resulting from the use of coal, on a unit of energy basis, then it is likely that the use of photovoltaics entails public health consequences considerably smaller than those of coal, since sulfates and nitrous oxides, organic, and other trace materials from coal are already known to affect public health significantly. Where photovoltaic impacts are larger than the exactly analogous effects of coal combustion, it will be necessary to consider the relative importance of other categories of impacts.

Comparability

It is useful at the outset to identify some of the differences between coal and photovoltaic systems which must be taken into account in order to arrive at meaningful comparisons of impacts on public health. One difference is that the connections between emissons and health effects are different for coal and photovoltaic energy systems. A coal plant may vent effluents through a 1000-foot stack, resulting in a considerable dilution of harmful materials before populations are affected. If there are thresholds below which most individuals experience few or no adverse effects, this dilution can be extremely important. In contrast, photovoltaics manufacturing facilities are not likely to involve such dissipative measures (unless required by regulation)* and may be built in closer proximity to higher population densities (in part a consequence of the high labor intensity of the technology). For decentralized applications— residential, neighborhood or commercial—potential sources of accidental release, primarily fires, are necessarily close to high population densities. It is thus necessary not only to compare total emissions for coal and photovoltaics systems but also to model the appropriate transport mechanisms leading to human exposures.

---

*Recent regulatory law (Clean Air Act amendments of 1977), in fact, prohibits dispersive measures, such as stack height increases.

The second difference is one of scale. The scale at which coal combustion occurs is set, by economics, at about 1 GWe (1000 MWe), for a new plant, and the size of this plant determines the ambient levels of pollutants to which humans are exposed. Such a plant, operating at 70 percent capacity factor, produces 0.7 GWe-year of electricity for each year of operation, equivalent to perhaps 0.6 GWe-yr. delivered to point of use. The size of refining and manufacturing facilities for photovoltaics will similarly affect ambient levels. In comparing with coal, it is thus necessary to specify photovoltaic plant sizes. One could do this by designing economically optimal facilities and then pro-rating the health impacts over the total power generated by the photovoltaic devices produced in the plants over the lifetimes of these devices. Alternatively, one could assume a manufacturing plant size which, in steady state operation, was responsible for the same amount of power generation as the model coal plant. For manufacturing facilities, both procedures appear to give comparable plant sizes: design studies (e.g., Lockheed[1]) suggest PV manufacturing facilities with production capacities of about 100 peak megawatts of PV capacity per year. The alternative calculation gives a plant size of 150-300 MWe peak PV capacity produced per year.[2] We shall use a plant size of 200 MWe peak/year since it is more directly comparable with coal and nuclear power plants.

In the case of cadmium and gallium arsenide devices, public health effects may occur not only at the raw material, refining, and manufacturing states, but also after the devices are installed and in use. It is also necessary, therefore, to establish equivalences between small, intermediate, and large-scale PV applications and the standard coal plant. Central station photovoltaic systems are obviously the most directly comparable to coal plants (except that PV plants might be more remotely located). For residential PV systems of 5 kilowatt peak capacity, approximately 600,000 installations are required to supply the same electricity as a 1 GWe coal plant (allowing about 15 percent loss for transmission and other losses for the coal plant). Similarly, approximately six thousand 500 kw intermediate-scale PV installations are, in total, equivalent to a coal plant.

Public health impacts from cadmium and gallium arsenide installations appear to arise primarily from fires, especially in intermediate and small-scale applications. While it is very unlikely that a fire involving all of a properly designed central station PV installation could occur, fires involving the roofs of houses and commercial establishments are fairly common. In order to compute the effects on public health of such fires on a unit of energy basis, it is necessary to know the rates of incidence of roof fires and the influence photovoltaic systems would have on this rate and on the nature of the fires occurring.

According to available fire statistics and estimates (discussed in detail below), the 600,000 residential installations, equivalent in total electrical production to a coal plant, would suffer about 200 fires involving arrays each year, while six thousand 500 kw installations could experience on the order of one fire per year. Like the emissions associated with production, the emissions from these events can be modeled and compared with the continuous emission from the coal plant, similarly modeled.

Assessment of Public Health Impacts

The determination of releases, the modeling of atmospheric transport and human exposures, and the estimation of the resulting public health effects are all difficult tasks. In order to estimate impacts, it is necessary to know or assume:

- the quantities of potentially harmful material present at each stage of manufacture or application
- release fraction—the fraction of such materials which are released, routinely at each process step or accidentally in fires or other events
- the precise chemical composition of the material released. The compounds released may be altered by the mechanism of release or by other processes (for example, cadmium sulfide may be oxidized in a fire). Different compounds of the same element can have different health effects
- the nature of the release—whether at ground level or from the top of a roof or exhaust stack, whether hot or cold. A hot release will rise, expanding adiabatically until cooled, effectively increasing the height of release. The height of the release affects the maximum concentration experienced at ground level
- atmospheric transport processes affecting the magnitude and duration of human exposures
- population distribution relative to distribution of pollutants
- the nature and effect of exposures to various compounds: inhalation versus ingestion, retention and lifetime in the body, acute versus cumulative effects, toxic versus carcinogenic impacts, potentiating and antagonizing synergisms and so forth

The first three entries above are specific to the photovoltaic technology involved; each will be discussed and compared to coal in subsequent sections. For the purpose of rough comparison between photovoltaics and coal, the last (and difficult) stage of analysis can be ignored in situations in which qualitatively similar exposures can be expected. For example, lung uptake of similar amounts of cadmium oxide from the two sources will have comparable health impacts. In the context of our comparisons, however, we do attempt when possible to characterize the kind and severity of effects that might occur, and to indicate where major uncertainties remain. Before turning to these comparisons, it is useful to consider the model for atmospheric transport and concentration under various release conditions used in this study. A more detailed description of the model is given in the appendix.

For the purposes of this study, a simple three-dimensional Gaussian diffusion model for the atmospheric dispersion of a general pollutant is employed. Given the conditions of release and general atmospheric characteristics, this model gives the average concentration of material at any point downwind of the source. The conditions which must be specified include effective height of release, average wind speed, rate of release and crosswind and vertical dispersion parameters. A more complex model would take account of deposition and the effects of surface irregularities, shielding, and so forth. However, this increased sophistication would introduce a level of detail (and many more assumptions) which is not justified by present knowledge or by the results of the simpler model; in general, significant exposures occur primarily under circumstances in which greater model complexity would make little difference in predicted exposure.

It should be kept in mind that the models and assumptions used in the analysis below are far from definitive. Our primary objective is to provide evaluations for basic policy purposes—giving only a rough measure of the relative hazards associated with different photovoltaic technologies and applications—and to identify areas where further research or improvements are needed. Refinements of our estimates will be required in the future in order to reflect changes in technology and in epidemiological knowledge. In addition, more detailed models must be used to evaluate the local impacts of particular facilities or installations.

The three sections that follow examine the potential public health impacts associated with the three generic photovoltaic technologies. For each, we estimate the release rates of hazardous materials at each step in production and use; summarize the results of our dispersion model for each release; assess the health impacts associated with the predicted potential public exposures; and draw comparisons with coal. (Detailed modeling results—including coal comparisons—are given in the appendix, along with discussion of the particular assumptions made in applying the model to specific cases.)

Silicon: Materials and Emissions

Despite experience in the semiconductor industry, the mass flows and process losses involved in silicon cell production are uncertain, with estimates in the literature varying widely. Often these differences can be traced to different assumptions about the efficiency of the process steps by which raw silicon is eventually turned into cells; however, there is rarely enough information to identify precisely where different assumptions are made. We shall use the description in fig. 5.1, which is based on the most extensively documented analysis available.[3] It should be noted that there are compelling economic reasons to alter this picture since it appears difficult to meet price goals with the technology illustrated. Changes in mass flows or processes will affect the magnitudes of various impacts.

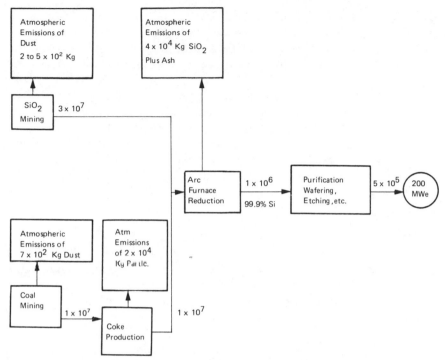

Material flows based on [2], assuming a cell thickness of 150 microns and a photovoltaic efficiency of 14%. All mass flows are in kilograms. Emission rate assumptions are discussed in text.

Fig. 5.1. Silicon process flows.

The largest potential health impacts associated with silicon technology occur early in the production process; there appear to be few significant public hazards subsequent to the production of metallurgical grade material. The primary hazards identified include:

- mining of silica and coal (for coke), resulting in the release of particulates
- production of coke from coal, resulting in the emission of suspended particulates, hydrocarbons, and other materials
- refining of silicon, resulting in the release of particulates of silicon dioxide and monoxide and fly ash.

The last of these is by far the most significant. The mining of silicon and coal for production of solar cells involves a volume of material only about 2 percent that of a coal power plant. Production of coke produces emissions qualitatively similar to those from coal burning; here, the volume of material involved is only about 1 percent that of the coal power plant. Refining emissions are of relatively greater importance since atmospheric release fractions are large and since a large fraction of the particulates emitted are very small (many less than one micron) and are most likely to remain suspended in the air, inhaled, and retained in the lung.

Respirable particulate emissions from the electric arc furnace (from silicon and coke) are about 100 grams per minute (for a plant producing material sufficient for 200 MWe capacity per year). Estimates of the fraction smaller than a few microns are uncertain but easily could exceed 50 percent. In contrast, our standard coal plant emits about 3 kilograms of respirable particulates per minute[4] (Western coal with 99 percent efficiency by weight in particulate recovery). The fraction in the micron range is probably smaller than for the electric arc furnace.

Silicon: Population Exposures and Health Effects

The particulate releases from coal and silicon refining were modeled to determine ground-level concentrations and resulting population exposures. The basic pattern found is that the larger volume of coal particulates is diluted and distributed over a much larger area than are the silicon particulates. Thus, the results of our model indicate that, given similar population densities, particulates from coal will affect approximately three orders of magnitude more people at any given concentration than a silicon refining plant responsible for the same electricity generation. For example, under neutral weather conditions, silicon refining could expose approximately 50,000 people to a particulate concentration of one microgram per cubic meter (1 $\mu g/m^3$, or 10^{-6} g/m^3); a coal plant could expose approximately 20,000,000 people to that same concentration of particulates. However, it should be remembered that this comparison is not precise: the composition of particulate emissions from silicon refining and coal combustion are different.* Since the dispersion model is rather idealized, the comparisons are thus only suggestive. What is perhaps most important is that silicon particulates represent only a small fraction of the public health impacts of coal, but are probably the major direct public impact of silicon photovoltaic technology. Thus, even comparability of particulate impacts would imply a significant superiority for silicon photovoltaics.

The comparative evaluation may, however, conceal the possibility of appreciable health impacts from both technologies. At present, there are not enough data available to convert concentration levels into public health consequences. The occupational standards discussed in chapter 4 are based on

---

*About 20 percent of coal plant particulate emissions are silicon. It is possible, of course, to consider *only* the silicon component of coal emissions: roughly one hundred thousand people may be exposed to 1 $\mu g/m^3$ of *silicon* particulates by a coal plant.

experience with worker populations exposed to workplace concentrations for limited periods of time. The current OSHA respirable silicon standard is $10^{-4} g/m^3$ (crystalline silica) for eight-hour shifts. It is possible that better studies will reveal more subtle health effects than established thus far, resulting in a reduction in the standard; this is especially likely for submicron particulates. There is as yet no federal standard for public exposure to respirable silicon. Public health standards, however, are normally set considerably lower—often about two orders of magnitude lower—than occupational standards since the sensitivity of certain groups (e.g., infants, the chronically ill, and the aged) will be much greater than that of healthy worker populations.[5] Thus it is reasonable to expect a public silicon standard of 10^{-6} g/m^3. But as noted above, concentrations produced by both coal plants and silicon refineries exceed this level for large numbers of people. (Indeed, under unstable weather conditions, exposures within an order of magnitude of the *occupational* standard are possible for hundreds of people.)

Cadmium: Materials and Emissions

A great many of variations on cadmium sulfide cells are possible. However, in order to perform an evaluation of public health impacts, it is necessary to establish flows of materials. We shall use the quantities shown in fig. 5.2, where two cell configurations are illustrated. In general, thinner film cells are expected to have lower photovoltaic efficiency and be accompanied by higher losses in manufacturing. As a result, the amount of cadmium required for a given photovoltaic capacity probably will not vary greatly with design. Figure 5.2 also shows estimated atmospheric emission rates at each step. These

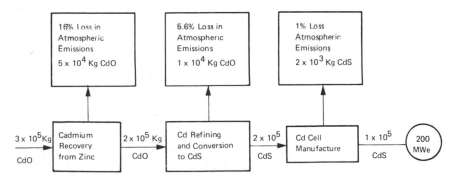

Material flows based on cells 6 microns thick with 6% photovoltaic efficiency. Process efficiency in manufacturing is assumed to be 50% by weight. All mass flows are in kilograms. Emission rate assumptions are discussed in text.

Fig. 5.2. Cadmium process flows.

estimates are based on present levels of control; loss rates depend on avail-ability and commercial utilization of control technology and on regulatory restrictions and great improvements are possible. The largest releases appear to occur during primary zinc refining; a common estimate is for a loss rate, in the United States, of 15 to 16 percent (by weight) of cadmium present in input zinc ore.[6] According to this estimate, about 50 metric tons* of cadmium would be released for each 200 MWe of PV capacity ultimately produced. The accuracy of the 15 to 16 percent estimate is open to some question since it is based on national mass balances. Empirical data on actual stack emissions appear to be lacking, with measurements complicated by institutional and technical problems. Significant amounts of cadmium may be lost as vapor or as exceedingly fine particulates—both difficult to measure, and to control. Estimates of recovery rates achieved with the best present technology appear to range from 90 percent to about 99 percent (reviewed in Fulkerson[9]). The smallest losses appear to occur in electrolytic zinc recovery facilities; these are used primarily abroad. Since impacts are potentially significant, it is important that uncertainties about present releases be resolved and that opportunities for reductions be explored. We shall discuss below the reasons for our assumption that the release during zinc refining should be attributed to use of photovoltaics.

Following this initial step, the cadmium is purified and reacted with sulfur to make cadmium sulfide. The atmospheric release fraction at this stage is quoted as being about 5 percent** in the form of CdO, or about 14 metric tons per 200 MWe peak capacity. The possibility of reducing this release will be discussed below.

Cadmium can also be released during subsequent processing and transport steps; however, the release fractions are far less certain. Handling risks are present because the cadmium sulfide is in the form of a fine powder up until the final cell fabrication steps. While automation and the high value of the output product will justify much better control, an upper bound on releases probably can be derived from the paint industry where release fractions are as high as 1 percent. This is either a serious occupational problem or, more likely, a public hazard since forced ventilation is the probable solution to high

---

*This is much larger than the 0.6 metric ton estimated by Energy and Environmental Associates[7] and subsequently cited elsewhere (e.g., ERDA,[8]. The EEA report estimates emissions of "14 pounds per ton (33 kg per metric ton)"—not equivalent quantities—citing an EPA report of 1975.[6] The citation appears incorrect: the EPA report (p. 5–7) lists an atmospheric loss in zinc operations of 155 kg cadmium per metric ton of cadmium input in ore. The EPA figure is based on an Oak Ridge report[9] which in turn cites a study by W.E. Davis and Associates.[10] The EPA report correctly represents the original analysis. However, the EEA report, and subsequent reports based on it, appear to be in error.

**The loss rate is derived from the process diagrams. Figures 9a and b, detailed in Gandel [3], p. 52–3.

workplace concentrations. *If* releases were as high as in the paint industry, up to two metric tons of cadmium (as CdS) could be released for each 200 MWe of peak capacity produced. These would be primarily routine handling releases (fugitive dust). Because the amounts of cadmium involved are very large compared to the toxic dose, process steps should also be examined carefully for possible accident sequences.

Thus, for the three production stages, the releases from cadmium photovoltaic technology are qualitatively similar to those of silicon, consisting of continuous emissions from the relatively large industrial facilities involved in CdS cell production. These emissions are also those most directly comparable to those from coal plants. Indeed, one can compare the cadmium releases of up to 60 metric tons annually from photovoltaic manufacturing (for our 200 MWe annual cell output) with cadmium releases from coal combustion. The mean concentration of cadmium in coal is 2.52 ppm; with a release fraction of 0.35, this results in the emission of about two metric tons of cadmium (as CdO) per year for our model coal plant,* considerably less than for photovoltaics, under the assumption that refining emissions could be attributable to photovoltaics. Of course, this comparison is far from complete since coal combustion results in the release of other hazardous substances.

Unlike silicon cells, however, the use of cadmium sulfide cells involves risks arising from fires affecting installed arrays. Depending on the temperatures reached, it can be expected that fires involving rooftop installations will result in emission of a certain fraction of the cadmium present, as cadmium oxide or sulfide. This release fraction is unknown; since significant impacts are possible, empirical determinations of releases under various circumstances and for different configurations should be made before wide-scale use. For purposes of analysis, we shall assume a release fraction of 0.1. We believe this to be a conservative assumption—that is, that the actual release fraction will very likely be found to be smaller.

As discussed earlier, about 600,000 five-kilowatt residential installations are needed to produce an electrical output equivalent to that of a large coal plant. Each installation will contain 1-3 kg of cadmium; the 200 major fires to be expected annually may thus result in the release of 20 to 60 kg of cadmium. This quantity is small compared to that emitted in the production of cadmium and cadmium cells, or in the combustion of coal, and would be episodic rather than continuous. An individual would thus receive a one-time dose or a sequence of one-time doses of varying severity rather than a relatively

---

*The range of cadmium concentrations in coal is large—from 0.1 to 65.0 ppm; however, the standard deviation about the mean (2.52 ppm) is only 7.6.[11] In computing the annual coal plant release we assume 10,000 Btu/pound, 700 MWe-years/year output and 36 percent thermal efficiency. As a check, we compare with an EPA study[12] which reports 0-130 μ/m^3 for power plant stack gases with a flow rate of 100 m^3/MWe/min. This gives a cadmium emission of 0-4.8 metric tons per year.

continuous exposure (as in the case of coal or cadmium industrial facilities). However, it may still be possible for some individuals to receive significant doses since releases will occur in areas of high population density. It is thus necessary to model each type of release independently.

Cadmium: Population Exposures and Health Effects

As with silicon, we modeled cadmium releases—and resulting population exposures—from coal, from the three stages in the production of cadmium sulfide cells, and from house fires involving cadmium sulfide arrays. At each production stage, concentrations of cadmium due to photovoltaics are two to three orders of magnitude higher than for coal under all weather conditions and at all distances from the plant. The number of people exposed to any given level of Cd is one to two orders of magnitude larger for photovoltaics than for coal. By far the greatest exposure comes at the refining and conversion step. Unless measures to reduce, dissipate, or isolate emissions from this production step are taken, several thousand people could experience routinely a cadmium level of 10^{-6} g/m^3, and up to 100,000 (depending on weather) a level of 10^{-7} g/m^3. (By comparison, a coal plant might expose a few dozen people to a cadmium level of 10^{-6} g/m^3 and about 500 to a level of 10^{-7} g/m^3.) Only a few hundred people are exposed to comparable levels at the other two cadmium production stages.*

Cadmium releases due to house fires involving photovoltaic arrays are somewhat more difficult to compare with coal; however, their effect seems to be much smaller than effects from the production steps. From a single fire, few if any people will be exposed to levels between 10^{-4} and 10^{-6} g/m^3 for as long as ten minutes. While some thousands of people may be exposed to cadmium from a fire, the exposure will be either for very short times or at very low concentrations compared with cadmium-array production steps or even with coal.

It is important to ask whether these exposure levels could have a significant impact on public health. It is evident that there is little chance of prompt fatal exposures for most of the facilities discussed above, or for array fires. A prompt fatal exposure is believed to be of order 1 gram-minute/m^3[13] though it could be lower for sensitive individuals. Only under extreme worst case conditions, prevailing for some days, could individuals very near cadmium facilities receive such an exposure.** Exposures due to house fires are three or more orders of magnitude too small to result in fatal effects.

---

*However, this assumes the low population density of 100/mi^2 surrounding the zinc smelter. For the other facilities, including the coal plant, we assume population density of 1000/mi^2.

**The worst situation appears to occur for an individual living 100 meters directly downwind of a fabrication plant venting 1 percent of throughput at ground level. About

The more difficult health issue is whether chronic exposure to lower concentrations can have significant individual or collective societal health impacts. This could occur through buildup in specific human organs (the half-life of cadmium in the body appears to be 15 years or more), causing degradation of function or carcinogenic effects.

It is believed that renal cortex concentrations of 200 ppm can lead to kidney dysfunction in sensitive individuals. Possible effects include proteinuria and hypertension. Since about one-third of the body burden of cadmium deposits in the kidney and since lung absorption may be as high as 40 percent, inhalation exposure levels should be kept low enough to prevent buildup to damaging levels. For 40 percent retention, an average concentration of 0.8 $\mu g/m^3$ over a period of 50 years would be sufficient to attain a 200 ppm renal concentration.* However, there are already other sources of cadmium in the environment. Dietary intake in the U.S. averages 60 μg/day with up to 5 percent retention. Thus, average ambient concentrations should be kept below about 0.5 $\mu g/m^3$.** Presumably, a safety margin should be applied to this figure.***

While ambient concentrations of cadmium from smelters, refining and conversion facilities, and manufacturing plants can occasionally exceed the 0.5 $\mu g/m^3$ level noted above—under the pessimistic release assumptions made here—it is very unlikely that long-term *average* concentrations experienced by individuals would be so high. Array fires would have even smaller effects; according to our model, the maximum individual uptake resulting from a house fire would be about 6 μg. This is equivalent to the uptake resulting from smoking about eight packs of cigarettes and does not appear to pose a significant individual risk. All of these conclusions are, of course, subject to verification of dose-response relationships which are at present somewhat conjectural. In particular, it is important that epidemiological studies be conducted in those areas in which cadmium concentrations are already high. It is possible that more subtle health effects are connected with ambient concentrations lower than those now thought hazardous.****

---

one week would be required for a fatal exposure; however, none of these conditions is very likely to exist.

*This is based on the calculation of Friberg,[14] modified to account for a possibly higher retention factor.

**This is about a factor of 100 below the occupational standard for eight-hour daily exposures.

***Smokers and those in the vicinity of smokers are already exposed to relatively high levels of cadmium and thus may be at greater risk. For unknown reasons, cigarettes contain 1 to 2 μg cadmium each; about 10 percent of this is inhaled and 40 percent released to the local environment. This is equivalent, for a pack-a-day smoker, to an average of 1-2 $\mu g/m^3$ in the air. Smokers may thus already exceed harmful levels of cadmium intake, a possibility which may also correlate with their higher incidence of hypertension.

****An example is lead, which is now suspected of being responsible for developmental and learning problems in children at levels previously thought harmless.

The final area of potential concern is the possibility that cadmium compounds could play a role in carcinogenesis. If this were the case, health assessments of cadmium photovoltaic systems would have to deal with a much more difficult problem since widescale low levels of exposures could be responsible—through the stochastic processes characterizing cancer induction—for significant numbers of health effects. The evidence of carcinogenicity for cadmium compounds is increasing, with greater than normal, but yet statistically inconclusive, incidence of cancers of the genito-urinary system reported for some occupational groups. However, the lack of conclusive data even for highly exposed groups is encouraging, implying that the ratio of total cancers to total human dose is relatively small.

For purposes of later analysis it may be useful to indicate the total population dose—the effective human uptake—which could occur under various weather conditions and for different facilities. These are shown in table 5.1.

Table 5.1. Human Cadmium Uptake.
(per GWe-year)

Facility/Condition	Maximum Population Dose (1)
Smelter (population density = $100/mi^2$)	
Unstable with inversion	10 person-grams
Neutral, 4 mph wind	3 person-grams
Refining and Conversion (pop. = $1000/mi^2$)	
Unstable with inversion	6 person-grams
Neutral	2 person-grams
Cell and Array Fabrication ($1000/mi^2$)	
Neutral	4 person-grams (3)
Array (4) Fires ($5000/mi^2$)	0.01–1.0 person-gram
Coal Plant ($1000/mi^2$)	
Cadmium	1 person-gram

(1) Assumes breathing rate of 10 m^3/day and retention of 40 percent of inhaled cadmium; figures should be considered only approximate and dependent upon model assumptions. Substantial contributions come from large numbers of people exposed to very low concentrations. We have somewhat arbitrarily cut off the sums over populations at about 500,000 people. Small but significant numbers of additional people may experience health effects even at extremely low concentrations. A more sophisticated model would be required to sum over such effects properly; the difference in result would be no more than a few percent.

(2) While present facilities may release 5 percent of throughput, we have assumed significant control efforts to reduce the release fraction to 1 percent. This is then consistent with the assumptions made subsequently for arsenic.

(3) Assumes 1 percent loss of cadmium throughput, a high upper bound.

(4) Assumes 200 fires affecting 5-kw residential installations.

The quantities shown are very rough upper bounds, with estimates based on pessimistic assumptions about cadmium release rates, lack of shielding, deposition, depletion, and so forth; they should thus only be used to form a preliminary impression of worst case conditions. It should also be noted that this way of assessing health impacts is useful only if there is no threshold for cadmium-induced health effects, since a considerable part of each total results from the exposure of very large numbers of people to very small doses. If there is no threshold—for carcinogenesis, hypertension or other effects—then collective human uptake can be used to form estimates of total health impacts. This would be similar to the situation with radiation-induced disease, where a total population exposure can be thought of as being responsible for a certain number of cancers or mutagenic events,* even if these events cannot be correlated empirically with the exposure of particular individuals. At present, it is not possible to make this kind of estimate for the effects of cadmium exposures.

Gallium Arsenide: Materials and Emissions

Two different lines of gallium arsenide technology have been proposed: thin polycrystalline films—much like those envisioned for cadmium sulfide—and thicker monocrystalline cells to be used under Fresnel lenses or other concentrating devices. It is generally believed that concentrating arrays would be used primarily in intermediate scale and central station applications; the thin films may allow residential applications. The primary source of public health impacts appears to arise from possible releases of inorganic arsenic.** Potential releases of arsenic occur at all stages of material processing and fabrication. The flows and losses of material assumed for the production of both types of cells are shown in fig. 5.3. All flows are normalized to the production of 200 MWe-peak of capacity—the amount necessary to sustain an average electrical output equal to that of a conventional central station plant.

Arsenic is present in most copper, lead, and zinc ores and being highly volatile (vaporizing at 169° C), is released when these primary metals are recovered by smelting or other means. A substantial but uncertain fraction of the arsenic condenses in plant exhausts, and fine particulates of arsenic (as the trioxide) are (partially) captured by electrostatic precipitators, baghouses, or other control devices. The resulting flue dust (a mixture of copper, arsenic, and other material) can then be processed to separate out the arsenic. Release

---

*A frequently cited measure of radiation effects correlates one cancer (and about one multifactorial genetic event) with a total exposure of 10,000 man-rem. A similar equation for cadmium-induced effects would correlate total human uptake (in grams, say) with occurrence of particular health effects.

**Inorganic arsenic compounds appear, from epidemiological studies, to be of much greater concern in regard to health than organic compounds; the latter are used predominantly as herbicides and pesticides.

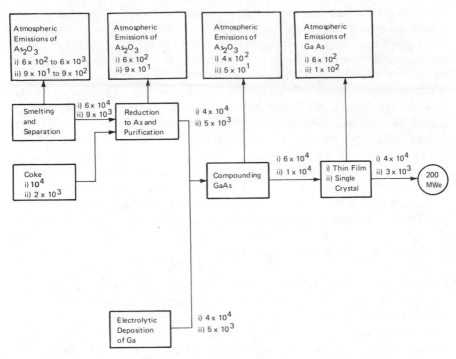

Thin film cells are assumed to be 5 microns thick and have a photovoltaic efficiency of 7%. Single crystal cells are 270 microns thick and have 17% photovoltaic efficiency at a concentration ratio of 500:1. Manufacturing process efficiency is assumed to be 30% by weight. All mass flows are in kilograms. Emission rate assumptions are discussed in text.

Fig. 5.3. Arsenic process flows.

fractions (both stack emissions and fugitive dust) at copper smelters appear to range from a few percent to 30 percent or higher.* Mass flow calculations suggest that the amount of As_2O_3 released could be as high as 25 percent of total domestic production (3,000 tons lost compared to 12,000 tons recovered), with most of this loss due to uncontrolled emissions at some smelters.**

---

*Evaluations in this section are based on information provided by the Environmental Protection Agency[15] and from Sullivan.[16]

**The present domestic market for arsenic is about 30,000 tons (As_2O_3) per year. A great deal of arsenic is imported despite the relatively large losses at domestic facilities; reduction of these emissions thus appears to depend more on regulatory than economic imperatives. As a result, one would expect that increased demand for inorganic arsenic for use in photovoltaic systems would have minimal effect on incentives to reduce emis-

At present, arsenic trioxide (about 12,000 tons annually) is produced commercially in the U.S. only by the ASARCO plant in Tacoma, Wash. Flue dusts from a number of United States and foreign copper smelters, as well as copper ores with especially high arsenic content, are processed at the Tacoma facility. Until recently, arsenic emissions at this plant were at or above the highest levels found at copper smelters. In the past few years, the Tacoma plant has been required to install new control devices for several phases of its operation. Success in controlling emissions is still uncertain; the very small particle size involved means that baghouses and electrostatic precipitators are less than fully efficient. For the purpose of our analysis, we assume a loss rate of 5 percent.* (For our dispersion model, we assume as well a population density of $100/mi^2$, considerably less than that of the Tacoma area. Siting considerations should be of first importance if photovoltaics should lead to the building of a new arsenic trioxide production plant.)

Emission rates for subsequent processing steps are essentially unknown. However, except for the final step, one can use rates for existing industries as a basis for preliminary estimates. Reduction of arsenic trioxide to metal, using coke, and its purification by distillation has many of the same problems with emissions as smelting of primary ore and is similar to the processing of cadmium. For the purposes of our model, we assume an emission rate of 1 percent of throughput. This assumes a relatively high level of control, to be expected in response to recent regulatory efforts. (Under present practices, perhaps 5 percent is lost.) On a materials balance basis, such losses are smaller than those presently associated with primary production of As_2O_3. However, releases may occur in greater proximity to populations and may have correspondingly greater effects on health. It is therefore necessary to model potential exposures from the reduction/purification stage separately.

Once arsenic is obtained and purified, it must be compounded with gallium. Emission rates for this step can be estimated from data on emissions from the 25 facilities which produce industrial inorganic arsenic compounds (mostly catalysts and reagents). These facilities processed about 5,000 tons of arsenic in 1974 and were estimated[15] to be responsible for emission of about 170 tons of arsenic—a 3.4 percent loss rate. It is likely that new regulatory standards

---

sions; however, it may provide a market for material collected under emission restrictions. If DOE goals for PV capacity additions (10–20 GWe peak in 2000) were met entirely with gallium arsenide cells, demand for arsenic trioxide would increase by 3,000 to 6,000 metric tons per year, a quantity equal to a 10–20 percent increase in current consumption (and also comparable to domestic processing losses).

*We do not consider as attributable to photovoltaics any arsenic emitted during the production at copper smelters of flue dust which is then processed at a facility like the Tacoma plant. Because of its high volatility, it is likely that arsenic will be emitted during copper smelting whether or not it is of commercial value. We will discuss below the dissimilar care of cadmium, which requires especially high refining temperatures and so is normally left as an impurity in zinc unless there is commercial demand for the cadmium.

and the construction of new facilities in response to PV demand would result in somewhat lower rates. We will thus again use a loss rate of 1 percent. (It should be noted that this assumes considerable efforts at control, requiring regulatory or economic incentives.)

Since most arsenic compounds are readily oxidized in the presence of heat, it is likely that most emissions during compounding are in the form of fine particulates of arsenic trioxide. Thus, the kind and amount of material released, as well as the resulting concentrations and exposures, are essentially the same at the compounding step as at the reduction/purification step and need not be modeled separately.

However, the emissions from the final production steps, cell manufacturing and array assembly, will be somewhat different since releases probably will be from ventilation systems and in the form of fugitive dust, occurring with little or no heat release and near ground level. It is also likely that at this stage releases will be in the form of gallium arsenide, whose health impacts, relative to other inorganic arsenic compounds, are unknown. As in the case of cadmium cell fabrication, industrial analogies are of little help in estimating emission rates for these stages. All that is possible, given the current state of technological development, is to indicate what *could* occur at a particular level of control. In applying our dispersion model to these final stages of GaAs cell manufacture, we assumed a loss of 1 percent of arsenic throughput.

It is possible to compare the arsenic releases that result from photovoltaic manufacturing with arsenic releases from coal combustion. The mean concentration of arsenic in coal is 14 ppm; with a release fraction of 0.35, this results in the emission of about 12 metric tons of arsenic (as As_2O_3) per year for our model coal plant.* Photovoltaic materials production and manufacturing (for our 200 MWe cell output) result in the emission of about 4.5 metric tons of arsenic (90 percent as As_2O_3 and 10 percent as GaAs). Of course, public health effects depend not only on raw amounts emitted, but also on dispersion patterns and population densities near facilities.

As with CdS, gallium arsenide arrays also present a potential public-health hazard in use, in the case of house fires. Because arsenic vaporizes at a relatively low temperature, we assumed a roof fire would result in the release of 70 percent of the arsenic (as As_2O_3) in a residential array, or about 180 kg per year. Because fires result in episodic exposures, as discussed above in the case of cadmium sulfide cells, it is necessary to compute total exposure rather than continuous exposure levels.

In applying our dispersion model, we considered only the case of thin-film GaAs cells. To produce a given electrical output, single-crystal GaAs cells require considerably less material, and therefore should normally result in lower

---

*In computing the annual coal plant release we assume 10,000 Btu/pound, 700 MWe-years/year output, and 36 percent thermal efficiency.

emissions and exposure levels than those associated with thin-film cells. However, two important exceptions should be noted: first, the mechanical operations associated with fabrication of single-crystal wafers may prove more problematic in terms of emission control than the chemical process line approach used for thin films; second, single-crystal cells will most likely be appropriate only for central-station use with concentrators. While this eliminates the danger from house fires, such arrays will require constant cooling. A loss-of-coolant accident could conceivably result in the release of significant quantitires of arsenic—one of the few imaginable situations in which solar technology might involve risk of sudden large-scale accidents in operation. Designs for facilities to produce or employ single-crystal GaAs cells should take account of these potential hazards.

Gallium Arsenide: Population Exposures and Health Effects

As with the other generic photovoltaic technologies, we modeled arsenic releases from coal, from four stages in the production of GaAs cells, and from house fires involving GaAs arrays. In comparing the first three production stages with coal, a pattern emerges qualitatively similar to that for silicon. Coal burning releases more material, but at a greater height. The greater resulting dispersion and dilution means that maximum atmospheric arsenic levels are roughly comparable for coal and photovoltaics, but many more people experience any given arsenic level for coal than for photovoltaics. Under neutral weather conditions, neither technology causes exposure to levels about 5×10^{-8} g/m^3 of arsenic. In unstable conditions, with an atmospheric inversion, a coal power plant could expose roughly 1000 people to an arsenic level of 10^{-7}/m^3; an arsenic production, purification, or compounding facility could expose about 100 people to this same level.

The emission and exposure pattern is somewhat different for gallium arsenide cell manufacture and array assembly, since both the temperature and the height of release will probably be much lower than those to be found at the earlier production steps. Therefore, effluents will be carried and diluted very little, and uniquely high concentrations of arsenic may occur close to the plant. Assuming a 1 percent loss rate, arsenic concentrations within one kilometer of the plant could range from 10^{-4} to 10^{-6} g/m^3. A few dozen people might be exposed to such concentrations. (In addition, such a plant could expose several hundred people to a concentration of 10^{-7} g/m^3, a number comparable to that for the earlier production steps and lower than for coal.) These maximum concentrations near the plant result in much higher exposures than for other processing steps and are higher than the maximum levels produced by coal, indicating the need for caution in process and plant design for cell manufacturing. To keep public exposure levels below 1 μg/m^3 even in the vicinity of the plant would require that arsenic throughput during cell manufacture be controlled to better than one part in ten thousand. (As noted above, this is more likely possible with a chemical process line approach to thin-film technology than with a mechanical wafer process for use with concentrators.) High temperature processes are of particular concern since control under such conditions is most difficult to achieve.

The final source of public exposure to arsenic—releases due to house fires involving arrays—may pose the most serious risk to public health, especially if research reveals that even low concentrations of arsenic (or brief exposure to higher concentrations) can increase cancer risk. Our model indicates that for a single house fire, thousands of people could be exposed to the equivalent of 10^{-6} g/m^3 of arsenic for several hours. The total annual inhalation of arsenic due to array fires ranges from about 10 percent to 50 percent of that due to coal combustion, for the same amount of electricity generated.

The very important question of whether there is a safe level of exposure to inorganic arsenic compounds has not been resolved for either occupational or public groups. The old OSHA standard was 500 μg/m^3; in January 1975, OSHA proposed a permissible exposure limit (eight-hour average) of 4 μg/m^3. The new standard[17] is 10 μg/m^3. In promulgating the standard, OSHA asserted that there is not enough evidence that there is a safe level of exposure to arsenic compounds, given their putative carcinogenicity, and that the new standard is based on feasibility of achievement rather than on known health effect thresholds.

While public exposure limits have not been set, it can be expected that they will be more restrictive than the OSHA requirements; it is also likely that possible new industrial sources of inorganic arsenic will be regarded with considerable caution by regulatory bodies. An impression of the magnitude of this problem for photovoltaics can be obtained from a review of existing ambient concentrations. Analysis of particulate samples from the 280 National Air Surveillance Network (NASN) sites in 1974 showed that only 73 sites exceeded the arsenic detection limit of 0.001 μg/m^3 (quarterly average). Of these, 11 had values between 0.02 and 0.17 μg/m^3; seven of these are in heavily industrialized regions and four are near nonferrous smelters. The average site near a smelter had average concentrations of 0.06 μg/m^3 while the nonsmelter sites had 0.01 μg/m^3. While epidemiological studies are not conclusive, there are suggestions (e.g., a 1975 National Cancer Institute study) that there is excess mortality from respiratory cancers in areas near copper, lead, and zinc smelters. Whether excess cancer incidence is due to arsenic is unclear; other industrial pollutants are always present. A number of new studies are under way and better data may soon be available.[15] If average concentrations as low as 10^{-8} g/m^3 (0.01 μg/m^3) are shown to entail significant health risks, it is likely that arsenic emission control will have to be effective to better than one part in one million of throughput.

For future analyses, once dose response data are known, we present, in table 5.2, approximate quantities of arsenic retained by affected populations due to photovoltaics and coal plant operation. The quantities shown (which assume a high retention figure of 40 percent of inhaled arsenic) should be considered upper bounds, perhaps by an order of magnitude. The quantities of arsenic retained are comparable for photovoltaics and coal.

These comparisons do not, however, imply that arsenic exposures due to photovoltaic (or other sources) are without health consequences, only that the impacts are comparable to those already occurring due to other sources. The primary hazard associated with low concentrations of arsenic is its potential for inducing cancer. Dose/response data are still unavailable, but the importance of such data for energy policy decisions should not be underestimated. An example—which it should be noted, is purely hypothetical—will illustrate. Suppose ambient concentrations near a smelter average 1.0 $\mu g/m^3$ and suppose that respiratory cancer mortality in the population so exposed is eventually shown to be twice the normal rate of 379 deaths per 10^6 persons per year. On this basis (and assuming an inhalation rate of 20 m^3/day and steady state conditions), the collective inhalation of five grams of arsenic by a large population could be correlated with one excess cancer death. If lower dose rates are proportionally effective in initiating cancer (that is, if there is a linear dose/response relationship), then it is possible to convert the total population exposures for coal and photovoltaics above to imputed cancer deaths. Though this logical chain is as yet unsupported by data, there is clearly

Table 5.2. Human Arsenic Uptake.
(per GWe-year)

Facility/Condition	Maximum Population Dose (1)
As$_2$O$_3$ Production Loss (Population density 100 mi^2) Unstable with Inversion	0.1 person-gram
Refining and Conversion (2) (Population density 1000 mi^2) Unstable with Inversion	1 person-gram
Cell and Array Fabrication (3) Neutral Conditions	1 person-gram
Array Fires (5000/mi^2, 200 fires)	1–5 person-grams
Coal plant Arsenic	10 person-grams

(1) Thin-film technology. Assumes breathing rate of 10 m^3/day and 40 percent retention; figures very approximate due to model dependencies.
(2) Assumes 1 percent loss of throughput.
(3) Assumes 1 percent loss of throughput; an upper bound.

reason to be concerned about connections between arsenic emissions from coal plants, or gallium arsenide PV technology, and human health impacts.

Public Health Effects: Assessment Issues

In the course of our analysis of the pubic health impacts of energy sources, several interesting and difficult issues arose relating to (1) assignment of costs, (2) interacting effects, and (3) the need to view old processes from new perspectives. We mention them here not only because of their relevance to the comparative assessment of photovoltaics and other energy technologies, but also because they seem typical of problems that tend to arise in the attempt to understand the effects on public well-being of any new, large-scale enterprise.

The question of the assignment of costs is obviously a vital one and can, in specific instances, be debatable. It is important not to assume automatically that a social cost associated with an activity is in fact *caused* by that activity; it is also important not to be easily seduced into believing the reverse.

For example, there is a question whether the large cadmium releases associated with smelter operation should be attributed to use of photovoltaics. It is often argued that zinc smelting operations would occur independently of cadmium requirements for photovoltaic systems. It is further argued that increased demand for cadmium would in fact provide economic incentives for better control of cadmium emissions—especially in lead and copper smelters, where 50 percent (lead) to 90 percent (copper) of the cadmium in the ore is released—resulting in a net social *benefit*. In principle, these arguments are valid. However, in practice there are economic and technical reasons to be skeptical.

First, control technology (baghouse filters and electrostatic precipitators) are already being used for primary cadmium recovery in zinc smelters; cadmium escaping is in the form of vapor or very small particles (generally less than a few microns) which are difficult to recover. Better recovery is possible but expensive; instead of increasing recovery rates at domestic smelters, industry has found it economical to import cadmium flue dusts from abroad. It is thus likely that improved domestic recovery will result only from stricter regulation of emissions and the invention of better recovery devices. One such device, currently under development, is a fluidized bed of particles which could absorb metallic vapors or trap small particulates when exhausts are passed through the bed.

Second, the demand for cadmium of a mature photovoltaics industry may exceed expected supply, even assuming near-perfect recovery. In the 1960s, U.S. cadmium consumption (end-uses) was between 4600 and 6000 metric tons[18]—a rate projected to increase to about 14,000 MT/year by 2000—used mostly in electroplating, plastics and pigments. A photovoltaic industry installing 10-20 GWe-peak per year in 2000 (the DOE goal) would be using about 10,000 to 20,000 MT annually, about equal to demand for other uses. Such quantities could not be produced simply by capturing cadmium being released through inadequate controls and might not even be produced as a by-product of normal production increases in other smelting operations, unless zinc, lead, and copper ores with high cadmium content were selectively chosen for smelting—precisely what is done in the case of arsenic, as noted above. Such high levels of consumption are thus likely to result in attributable negative impacts.

The question of assignment of cost has another, thornier, aspect: the assignment of responsibility. The law is at present far from clear-cut regarding the sort of delayed and probabilistic effects that may be associated, for example, with the release of substantial amounts of arsenic during a house fire. Unless homeowners and manufacturers feel assured that they are either legally absolved from responsibility or adequately covered by insurance, there may be reluctance to produce or purchase potentially hazardous devices; there may also be reluctance on the part of communities to permit their use.

Another kind of issue that arose at several points in our analysis is the interconnection of kinds of social costs which might at first appear to be independent. The possibility of trade-offs, intentional or unintentional, recurred frequently: a measure that seems to ameliorate some cost all too often simply transfers it to another sphere. Workplace ventilation may increase public exposures. Dispersive methods that lower the atmospheric concentrations of pollutants near population centers may distribute the hazardous material over wider areas, perhaps where food is grown. If we import arsenic trioxide rather than increasing domestic production, we may quite literally be exporting a health hazard, since foreign populations will suffer the ill effects associated with its production. Since, as noted below, the safest way to deal with some house fires is to vent them through the roof, firefighters may have to weigh personal hazard against increased cancer risk (of presently unknown magnitude) for a large number of people. No generalization is possible for cases like these: the lesson is simply that they are very common, and that "ameliorative" measures must be examined carefully for such trade-offs.

The issue of interconnection raises also the question of synergistic health effects. As noted in the previous chapter, very little is known about what substances promote or retard the adverse effects of other substances. Better information in this area is essential for protecting public, as well as occupational, health. Better knowledge is also needed about which population subgroups have special susceptibilities to hazardous substances, particularly if the present trend in occupational regulation—keeping exposures low enough to protect all workers, not just average, health, unallergic, unpregnant young adults—is extended to the public-health sphere. A particular difficulty may arise if an increased susceptibility is arguably voluntary: society may have to decide whether it will pay an economic price to protect smokers—or the families and coworkers of smokers—from small increases in atmospheric cadmium.

The third interesting general issue our analysis raised is that when a new technology impinges upon a society, it may raise new questions and uncertainties about events and processes that were formerly thought well-understood. We realized this when we attempted to estimate cadmium and arsenic emissions due to house fires, and discovered that national fire statistics are not broken down in a way which allows very precise determination of the number of fires in which roof involvement occurs (fires are usually classified by origin, by the number of alarms or the amount of equipment involved, or by dollar loss). For example, it is known that fire departments responded to about 400,000 residential fires in 1974,*[19] but it is not known how many had roof involvement. If one assumes that between 1 and 10 percent had roof involvement then between 4000 and 40,000 roofs were affected by fire, or one roof for every 5000 to 50,000 people.

To refine an estimate of roof fire incidence, we interviewed a number of urban and suburban fire officials, finding a wide variation in annual rates— from about 1 roof per 2000 persons in the population to about 1 per 50,000. In general, the higher rates occurred in urban areas or in smaller industrialized cities while the lower rates occurred in moderate to high income residential suburbs. However, it also became clear that firefighting practices had some bearing on the range displayed. In many cases, firefighters will open holes in a roof to vent heat and gases, allowing entry of the building and preventing a "flashover," or spontaneous ignition of a larger portion of the interior. Fire officials were strong in their conviction that shaping a fire into a vertical configuration, by venting through the roof, saved lives and reduced property damage. However, this practice also led to a higher incidence of roof involvement in the fire. It is likely that the high incidence of fires involving the roof (1 per 2000 persons per year) in some areas is due to inclusion of fire events in which roof involvement results from the way in which the fire is fought.

This raises the possibility that the installation of photovoltaic panels on roofs may require changes in firefighting practices—if this is possible. While panels would generally cover only half the roof, allowing holes to be made in the other half, it is likely that the presence of valuable roof installations (and in the case of cadmium and gallium arsenide, the possibility of toxic releases) may inhibit firefighting measures which increase the risks of roof fires.** The trade-offs involved should be examined in much greater detail.

---

*About ten times as many fires actually occurred.

**This, in turn, raises the separate question of whether the addition of photovoltaic systems would increase or decrease the likelihood of fires. It is possible that electrical systems would be upgraded if PV systems are installed; on the other hand, new electrical hazards could be introduced. At present, about 8 percent of fires are of electrical origin.

It is clearly important to be alert for other areas in which unconventional technologies may themselves act to render unconventional any number of habitual activities.

Overview of Direct Public Health Effects

In this chapter, we have reviewed the nature and approximate magnitude of direct public health hazards which might arise in connection with the processing and use of the materials essential to photovoltaic systems: silicon, cadmium, and arsenic compounds. While a complete assessment also requires evaluation of the longer-term environmental roles played by these compounds *and* inclusion of the indirect health and environmental impacts which result from the use of other materials (such as aluminum or copper) and energy in making photovoltaic systems, it is useful here to review the relative importance of the direct impacts discussed in this chapter.

The most important observation is that the direct public health impacts of photovoltaics appear to be smaller than those of coal. In the case of silicon, particulate releases from coal affect large numbers of people at higher concentrations than do silicon refining emissions. While there are significant differences between the particulate emissions from coal and silicon refining, it is very unlikely that the mix of hydrocarbons, silicon, and other trace materials in coal particulates could be less harmful than the crystalline silica released in refining. In addition, the public health impacts of sulfates produced as a result of coal combustion may greatly exceed the effects of particulate emissions from coal or silicon photovoltaics.

The comparison is much more complex in the case of cadmium and arsenic-based technologies. Unlike silicon, where the high temperature processes which can result in the release of dangerous materials are present only at the refining stage, significant emissions of toxic and carcinogenic materials are possible at all points of manufacture and use of cadmium and arsenic photovoltaic systems. It is thus necessary to include a number of possible effects in comparing with coal.

While immediate toxic effects are unlikely as a result of cadmium releases, there is a possibility of significant long-term health impacts due to impairment of kidney function (leading to a number of effects, including hypertension) and carcinogenesis. Persistent exposure to cadmium concentrations in excess of about 1 $\mu g/m^3$ over a period of decades may result in a relatively high level of individual risk of kidney problems and hypertension. Under present control regimes, such levels would appear to be exceeded for many people (some thousands) due to releases from Cd refining and conversion (CdS) facilities and photovoltaic manufacturing facilities. Zinc smelter releases of cadmium are not the largest problem. Individual risk from cadmium-induced cancer is very small; however, the net impact in large populations may be significant.

Total uptake, by a population of about 500,000, is on order 10 grams per GWe-year of photovoltaic power generated. Average dose-response relationships are unknown and thus it is at present very difficult to estimate the effects of such a collective dose. House fires involving cadmium arrays present smaller individual and collective risks than do manufacturing steps.

Cadmium concentrations from coal combustion do not exceed 0.1 $\mu g/m^3$ and total population uptake is less than 1 gram. Thus, in this very limited category, coal may have lower consequences. However, other emissions of trace elements (including arsenic), polycyclic hydrocarbons and oxides of sulfur and nitrogen probably have much greater impact on human health (perhaps of order 100 premature deaths per Gwe-year, according to several epidemiological estimates).[20] It would be useful to have an epidemiological upper bound of cadmium-induced hypertension and cancer; we believe that such a bound would demonstrate the net adverse direct public health impact of cadmium photovoltaics to be smaller than that of coal.

The situation is somewhat different with arsenic. Here, the health issue is definitely carcinogenesis, and array fires are the dominant source of public exposures. However, population arsenic uptake from coal combustion—while delivered through relatively low exposure levels—exceeds that of gallium arsenide photovoltaics. This clearly establishes the superiority of photovoltaics in direct public health impacts, though the high volatility of arsenic makes array fires a source of some concern.

It is interesting to ask whether the above analysis provides a basis for choosing between photovoltaic suboptions. Both the number of risk centers and the nature of potential adverse public-health impacts seem to us to give grounds for some preference for silicon-based technology. With silicon photovoltaics there is only one significant source of emissions, the arc furnace, and silica does not appear to present cancer risks. However, firm grounds for quantitative comparison of the public health effects of different photovoltaic technologies do not yet exist. The comparison may change as the technologies evolve and as better information becomes available regarding the dose-response relationships for all the hazardous substances involved.

Perhaps more important issues in the near-term comparison of photovoltaic suboptions are the extent to which releases can be controlled and whether regulation will accomplish such control in ways which do not inhibit development and deployment of the technologies. In this regard, cadmium and arsenic technologies present problems at more points and are also potentially more vulnerable to regulatory change. As noted above, emissions from the silicon photovoltaic cycle are predominantly at one point; moreover, silica has been subject to occupational and environmental regulation for some time. While standards changes are possible, especially for small crystalline particulates, the control problem is focused on one manufacturing step. In addition, silica regulation does not have to deal with the potential for continuous ratcheting that accompanies identified carcinogens.

Cadmium and arsenic technologies present a series of control problems and opportunities for regulatory intervention. Moreover, the recent severe reduction (a factor of 50) in occupational exposures to inorganic arsenic—and the likelihood of a similar reduction for cadmium—may soon be followed by promulgation of regulations governing releases to the environment. The Clean Air Act amendments of 1977 require that EPA examine cadmium and arsenic emissions (not now subject to regulation) for public health hazards, and establish regulations governing these emissions if they will "cause, or contribute to, air pollution which may reasonably be anticipated to endanger public health." The regulatory prospect is not encouraging. The simplest solution to occupational exposures is often an increase in ventilation. Ratcheting on occupational standards may thus increase releases to the environment. However, it is likely that such emissions will in turn be regulated by EPA. Finally, the new Clean Air Act amendments prohibit dispersion techniques that "exceed good engineering practice," such as increasing stack heights, and insure that new sources will be the most severely restricted. These considerations imply an urgent need to anticipate regulatory problems as part of the technological evolution of cadmium and arsenic technologies.

REFERENCES

1. Sears, R.D. and Bradley, W.G., *Environmental Impact Statement for a Hypothetical 1000 MWe Photovoltaic Solar Electric Plant.* Prepared for the Environmental Protection Agency, Contract No. 68-02-1331. Cincinnati: Lockheed Missiles and Space Company, Inc., August 1977.
2. At 0.2 capacity factor, and a life of 10 to 20 years, each 100 MWe of peak PV capacity will, over its life, produce 200 to 400 MWe-years of electricity. Thus, a manufacturing plant of 150–300 MWe production capacity will, on the average, eventually be responsible for the generation of 0.6 GWe-year of power per year of operation, the same as the standard coal plant. Our assumption of 200 MWe peak capacity produced per year corresponds to an assumed lifetime of 15 years; that is, the PV capacity produced will generate the same 600 MWe-years of power over 15 years as the coal plant does in one year. In steady state operation, the PV manufacturing facility can be thought of as being equivalent to the coal plant.
3. Gandel, M.G. et al., *Assessment of Large-Scale Photovoltaic Materials Production.* Prepared for the Environmental Protection Agency, Contract No. EPA 68-02-1331. Cincinnati: Lockheed Missiles and Space Company, Inc., August 1977.
4. *Energy Alternatives: A Comparative Analysis.* The Science and Public Policy Program. Norman, OK: University of Oklahoma, May 1975.
5. The OSHA standard for respirable particulates generally is 5 mg/m^3 while the EPA standard (annual geometric mean) for public exposures is 75 μg/m^3, a factor of 70 lower. The maximum level allowed once a year is 260 μg/m^3

(24-hour average). It is interesting that maximum ambient concentrations from coal predicted by our model fall just within the annual standard.

6. *Scientific and Technical Assessment Report on Cadmium.* U.S. Environmental Protection Agency, Office of Research and Development, No. EPA-600/6-75-003. National Environmental Research Center, Research Triangle Park, NC, March 1975.

7. *Solar Program Assessment, Final Report, Photovoltaic Cells.* Prepared for the Environmental and Resource Assessments Branch, Energy Research and Development Administration by Energy and Environmental Analysis, Inc., Arlington, VA, January 7, 1967.

8. *Solar Program Assessment: Environmental Factors, Photovoltaics.* Environmental and Resource Assessments Branch, Energy Research and Development Administration, ERDA 77-47/3. Washington: March 1967.

9. Fulkerson, W., et al., *Cadmium—The Dissipated Element.* Oak Ridge National Laboratory, ORNL-NSF-EP-21. Oak Ridge, TN: January 1973.

10. *National Inventory of Sources and Emissions: Cadmium, Nickel and Asbestos: Cadmium, Section I.* W.E. Davis and Associates. Prepared for U.S. Department of Health, Education and Welfare, National Air Pollution Control Administration, Washington, D.C., under Contract No. CPA-22-69-131, February 1970.

11. Gruhl, J., et al., *Systematic Methodology for the Comparison of Environmental Control Technologies for Coal-Fired Generation.* MIT Energy Laboratory Report No. MIT-EL 76-012. Cambridge, MA: November 1976.

12. Dole, S.H. and Papetti, R.A., *Environmental Factors in the Production and Use of Energy.* Prepared for the Rockefeller Foundation, R-992-FF. Santa Monica, CA: RAND, January 1973.

13. *Criteria for a Recommended Standard . . . Occupational Exposure to Cadmium.* National Institute of Occupational Safety and Health. U.S. Department of Health, Education and Welfare, Pub. No. (NIOSH) 76-192. Washington: August 1976.

14. Friberg, L.M., et al., *Cadmium in the Environment.* Cleveland: Chemical Rubber Co. Press, 1971. As cited in *Cadmium—The Dissipated Element.* Oak Ridge National Laboratory, p. 267.

15. *Air Pollution Assessment Report on Arsenic.* Final Draft. Environmental Protection Agency, Strategies and Air Standards Division, Research Triangle Park, North Carolina, July 1976.

16. Sullivan, R.J., *Air Pollution Aspects of Arsenic and Its Compounds.* Litton Systems Inc., Environmental Systems Division. PH 22-68-25. Bethesda, MD: September 1969.

17. "Occupational Exposure to Inorganic Arsenic—Final Standard," *Federal Register.* Vol. 43, No. 88, May 5, 1978.

18. *1971 Minerals Yearbook, Vols. I–II.* U.S. Department of Interior, Bureau of Mines, Washington, DC, 1973.

19. *Highlights of the National Household Fire Survey.* U.S. Department of Commerce, National Fire Prevention and Control Administration. Washington: 1976.

20. Lave, L.B., and Seskin, E.P. *Air Pollution and Human Health*. Baltimore: Johns Hopkins University Press, published for Resources for the Future, 1977.

Chapter 6
Environmental Impacts

The impacts of energy technologies extend beyond direct exposures of workers and the public to hazardous substances: through heating effects or additions of moisture, energy facilities may affect local climate; CO_2 emitted in fossil fuel combustion may contribute to climatic change; the mining of energy materials and the siting of facilities may damage land and ecosystems; and pollutants emitted may enter water supplies and food chains, ultimately affecting human and other populations. The full list of such impacts is considerably longer and because of the complexity of natural systems, the true effects of energy generation and use are highly uncertain. However, relative comparisons between technologies are possible. It is generally believed that photovoltaic systems are relatively benign environmentally and this appears to be true, with possible exceptions. It is instructive to compare technologies in three categories of impact: land use, thermal and climatic effects, and emissions.

Land Use

Coal plants require the use of relatively small land areas for the actual plants and coal storage; however, over its lifetime, a coal plant may require the strip-mining of 3000 to 12,000 acres (about 5-18 mi^2).[1] While partial restoration is possible (and mandated by law), acid drainage and other problems may result from coal mining. In addition, an uncertain quantity of land will be needed for the disposal of fly ash and slurry recovered by scrubbers. A nuclear plant at first appears to use considerably less land than coal; mining and milling of uranium ore may directly affect 2100 to 3600 acres (at 0.2 percent ore concentration) or 3.3-5.6 mi^2, during the life of the plant. However, the airborne and aqueous emission of radon and its radioactive daughter products from mill tailings may effectively prohibit use of considerably larger areas for the thousands of years during which emissions continue.* Nuclear fuel waste disposal would deny additional areas to extensive human use for many centuries, though this area is likely to be small on a GWe-year basis since waste from a large number of reactors may be stored in a relatively small volume underground.

Central station solar facilities would occupy 10 to 40 square miles, depending on efficiency of conversion and spacing of arrays, to produce the same total annual electrical output as a coal or nuclear plant (700 MWe-years for the conventional plant, 3500 MWe at 0.2 average capacity factor for the solar plant); thus, solar electricity generation affects an area not a great deal larger than that committed to a coal plant over its lifetime. Decentralized photovoltaic systems would be placed largely on buildings, with no net impact on land use (except perhaps for denying roof use for other purposes). It is difficult to see how impacts on land and local ecosystems could be worse, even for central station photovoltaic systems, than for coal strip mining. Indeed, solar installations probably cause less permanent damage. Whether solar is preferable to nuclear power in this respect depends in part on the values attached to different effects on land and land use. For example, areas denied to human use as a precaution in case of leakage from sealed nuclear waste containers might sustain little if any aesthetic or ecological damage, whereas a central station solar facility would be highly visible, and relatively inhospitable to plants and animals. If it does not prove feasible to recycle solar arrays, their disposal may also have an impact on land use. Solid waste (under the assumptions about materials explained in chapter 7 and assuming a 20-year array lifetime) will amount to more than 12,000 m^3 per GWe-year. For cadmium- and arsenic-based arrays (as for nuclear waste), disposal sites would have to be remote, geologically stable, inaccessible to ground water, and secured from human entry, essentially forever. Such precautions are unnecessary in the case of silicon.

Central station generating plants—whether coal, nuclear or solar—can also affect land use through the need to establish high-voltage transmission corridors connecting generation centers with major load centers. For coal or nuclear plants, the distances currently involved are seldom more than 100 miles; however, lack of sites near load centers or health and safety concerns may eventually force more remote siting. Long-distance power lines are now used mostly for tying regional grids together or for transporting inexpensive hydroelectric power to large load centers (as is the case with transmission of electricity from the Pacific Northwest to California). For moderate blocks of power—say, less than 1 GWe—long-distance transmission is very expensive. However, for four or five or more GWe, unit transmission costs are reduced considerably, as shown

---

*The most recent proposals for dealing with tailings include denial of habitation in mining and milling districts and the reburial of tailings in exhausted mines.

in fig. 6.1. New transmission technologies, such as superconducting lines, may allow movement of large blocks of power at even lower unit cost.

This raises the possibility that very large solar collecting installations could be located far from load centers, perhaps in the desert Southwest. Desert siting would increase the effectiveness of solar collection since average insolation levels would be higher and since lower atmospheric moisture content would mean less absorption and scattering the photons in the wavelength regions important in photovoltaic generation. The land area needed for transmission corridors would be about 1 mi^2 per 10 miles length. For a large installation 1000 miles from the point of use, about 10-20 mi^2 of land use for transportation is attributable to that fraction of the photovoltaic array which is equivalent in total average power output to our standard coal or nuclear plant. Whether such an arrangement is desirable depends on further study of the trade-offs between transmission losses, economics, grid stability, and the advantages of remote siting. Again, decentralized arrays would avoid transmission impacts (and reduce the size and complexity of terminal distributional networks), though perhaps at some loss in economic or physical efficiency.

In the case of stand-alone photovoltaic installations, land may also be required for storage of solar-generated electricity. Presumably, battery or mechanical storage could be co-located with generation, with little net increase in land use (but perhaps with more severe impact). However, pumped hydro or some other storage systems for central station solar electricity would involve more significant use of land.

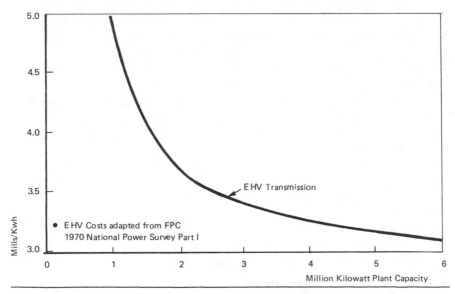

Fig. 6.1. Power transmission cost as a function of load carried.

Thermal and Climatic Effects

Coal and nuclear plants result in the emission of about two units of energy as heat for every unit of electricity generated (nuclear actually produces about 20 percent more heat than does coal). Heat from coal plants is in part dissipated directly into the atmosphere along with flue gases, with the remainder dissipated in cooling water. Heat from nuclear plants is dissipated primarily through water cooling. For once-through cooling of a nuclear plant, about 1000 cubic feet per second (cfs) are required (with a $10°$ C temperature rise); for evaporative cooling in cooling towers, about 30 cfs are required. Evaporative cooling lessens thermal impacts on rivers or lakes but at the expense of injecting water vapor (and dissolved minerals) into the atmosphere. The total amount of heat released directly by a large conventional power plant averages about 1.5×10^6 kilowatts, equivalent to the sun's heating of about 20-30 km^2 of land.* While such a heat release would have little effect on a global scale, local effects on climate may be more substantial, particularly if many coal or nuclear plants are co-located. Particulate emissions from coal combustion may also affect the earth's albedo (reflectivity).** Depending upon size and distribution, particulates can cause increased reflection of sunlight away from the earth, leading to global cooling, or trap infrared and reflected radiation, leading to global warming. The net effect is still unknown.

The effect of central station solar plants on local heat balances will probably be less than those of conventional plants. Solar panels will reduce reflectivity, potentially increasing local heating, but this will be at least partially compensated for by the conversion of some solar energy to electricity and transmission away from the site. Thus, PV systems will have no effect on local heating if a reduction by 10 percent of total solar flux in the amount of energy reflected back into space is compensated for by the 10 percent or so of sunlight converted to electricity. It would be useful to evaluate prototype installations in this context. In any event, it is very unlikely that net effects on the local heat balance could be as large as those due to the even more concentrated heat sources of coal or nuclear plants.*** Photovoltaic installations will have little or no effect on local water supplies, except during construction, and will do little to alter atmospheric moisture content.

---

*The yearly average solar insolation ranges from 0.16 to 0.26 kilowatts (1200–2000 $Btu/ft^2/day$). Of this, about half, on average, is absorbed and re-radiated (most within hours) as heat.

**About 30 to 50 percent of sunlight is reflected directly back into space on the average, a rate which is much higher in desert areas.

***The effect on local heat balance will be even less for decentralized use of arrays. The average house roof absorbs considerably more sunlight than does desert sand; a roof's reflectivity will not be appreciably lessened by the addition of solar panels.

Solar photovoltaics would help alleviate what may be an increasingly important effect of fossil fuel combustion: the buildup of atmospheric CO_2 and resulting effects on global climate. Carbon dioxide is transparent to most incoming solar radiation but not to reradiated infrared wavelengths; the result of increasing CO_2 concentrations is thus global warming. Due to deforestation, fossil fuels combustion, and other mechanisms, atmospheric CO_2 has increased by about 10 percent since the industrial revolution. Assuming that about one-third of the emitted CO_2 stays in the atmosphere (the rest being absorbed, primarily in oceans), the present rate of fossil combustion would add about 2 percent per decade to atmospheric CO_2. An increase in fossil combustion obviously would accelerate this process. It is believed, on theoretical grounds, that an increase of 20 percent in CO_2 content can cause an increase of $1°C$ in global temperature—an increase which undoubtedly would cause climatic change. However, many processes (including the addition of particulates from coal combustion) affect temperature and climatic changes. The net effect of fossil combustion and other factors on climate is thus uncertain. But the possibility that time and better data will reveal deleterious climatic change adds importance to efforts to develop new technologies which do not depend on releasing the carbon reservoirs in oil and coal built up by photosynthesis in millenia past.

Emissions

Significant environmental impacts may result from the addition to the environment of various substances from coal combustion, nuclear plant or fuel cycle operations, and from photovoltaic manufacturing and applications. The emissions of coal plants have been studied extensively, though there is still uncertainty about the ultimate environmental fate of these emissions. According to a 1973 Battelle study,[2] combustion of Eastern coal in a conventional boiler with wet limestone scrubbing results in the release of 3000 tons of particulates, 18,000 tons of nitrous oxides, 15,000 tons of sulfur oxides, and about 400 tons of hydrocarbons.* While some of the sulfur and nitrous oxides contribute to atmospheric particulate concentrations in the form of sulfates and nitrate byproducts of gaseous emissions, others increase the acidity of rainfall through the formation of nitric and sulfuric acids.

Airborne pollutants and acid rain can both affect ecosystems, to a degree and in ways which are not yet fully understood. The effect of acid rain is perhaps the most extensively documented, with strong evidence that pH reductions in soils and bodies of water are due to increased fossil fuel combustion. Sulfur dioxide appears to impede nitrogen fixation, necessary for plant growth, and

---

*Amounts are for emissions per 10^{12} Btus; we have converted to yearly plant output by using a heat input figure of 60×10^{12} Btu/yr.

to inhibit some of the processes involved in decomposition of dead organic matter. Combustion of coal—or its conversion to coke or liquid or gaseous fuels—at high temperatures (in excess of 400°C) results in the formation of potent carcinogens,[3] part of the hydrocarbon or particulate release, whose environmental fate is still uncertain.

A number of heavy metals and trace elements are present in coal, as shown in table 6.1. Release fractions in combustion are uncertain and depend on plant configuration and the method and efficiency of control. Under current regulatory standards, efforts are made to restrict emissions of sulfur and particulates: perhaps 90 percent of sulfur is removed and 99 percent (by weight) of particulates. Other emissions are restricted only to the extent that controls aimed at these components also work for other coal constitutents. Some evidence of the emission rates can be derived from examination of fly ash collected from effluent gases, as shown in table 6.1. However, the volatility of some elements, such as mercury, makes such determinations incomplete.

A very conservative basis for analysis is to assume that all metals are lost in atmospheric emissions. On this basis, annual mean-value emissions for a coal plant are: cadmium—5.8 tons; lead—80 tons; nickel—48 tons; arsenic—32 tons; vanadium—75 tons; and mercury—0.5 tons. Of course, controls (especially particulate removal) may reduce releases considerably.* The environmental fate of particular metals and metal compounds is a complicated issue, depending on chemical and physical form, resuspension rate, ion-exchange processes in soil, and so forth.

By comparison, manufacturing cadmium sulfide photovoltaic systems—using the assumptions in the preceding chapter—results in the release to the environment of up to 70 to 80 tons of cadmium per year (largely as CdO or CdS), while fires release on the order of 0.1 ton per year. Thus, manufacturing would be by far the dominant photovoltaics-related source (for present technologies) of additions of cadmium to the environment. Atmospheric arsenic emissions during manufacturing—also under the assumptions specified in chapter 5, and depending on the technology used—would total 200 to 6700 kg per year (with by far the greatest share—100–6000 kg, depending on the technology used and on the success of emission controls—resulting from As_2O_2 recovery from copper-smelting byproducts). Array fires might result in the release of 200 kg of arsenic per year. It can be expected that significant amounts of arsenic and cadmium will also enter the environment through waste water releases and treatment of solid wastes. All of these figures are for installations equivalent to our standard coal plant, which releases somewhat less cadmium and somewhat more arsenic than the respective photovoltaic systems but also much larger amounts of other metals and pollutants.

---

*It is occasionally useful to recall, too, that electricity generation—by any technology—is probably not the most socially costly part of the energy sector: the lead emission from coal is dwarfed by that from use of leaded gasoline.

Table 6.1. Constituents of Coal.

Constituent	Mean	Standard Deviation	Minimum	Maximum
As	14.02 ppm	17.70	0.50	93.00
B	102.21 ppm	54.65	5.00	224.00
Be	1.61 ppm	0.82	0.20	4.00
Br	15.42 ppm	5.92	4.00	52.00
Cd	2.52 ppm	7.60	0.10	65.00
Co	9.57 ppm	7.26	1.00	43.00
Cr	13.75 ppm	7.26	4.00	54.00
Cu	15.16 ppm	8.12	5.00	61.00
F	60.94 ppm	20.99	25.00	143.00
Ga	3.12 ppm	1.06	1.10	7.50
Ge	6.59 ppm	6.71	1.00	43.00
Hg	0.20 ppm	0.20	0.02	1.60
Mn	49.40 ppm	40.15	6.00	181.00
Mo	7.54 ppm	5.96	1.00	30.00
Ni	21.07 ppm	12.35	3.00	80.00
P	71.10 ppm	72.81	5.00	400.00
Pb	34.78 ppm	43.69	4.00	218.00
Sb	1.26 ppm	1.32	0.20	8.90
Se	2.08 ppm	1.10	0.45	7.70
Sn	4.79 ppm	6.15	1.00	51.00
V	32.71 ppm	12.03	11.00	78.00
Zn	272.29 ppm	694.23	6.00	5,350.00
Zr	72.46 ppm	57.78	8.00	133.00
Al	1.29%	0.45	0.43	3.04
Ca	0.77%	0.55	0.05	2.67
Cl	0.14%	0.14	0.01	0.54
Fe	1.92%	0.79	0.34	4.32
K	0.16%	0.06	0.02	0.43
Mg	0.05%	0.04	0.01	0.25
Na	0.05%	0.04	0.00	0.20
Si	2.49%	0.80	0.58	6.09
Ti	0.07%	0.02	0.02	0.15
Org. S	1.41%	0.65	0.31	3.09
Pyr. S	1.76%	0.86	0.06	3.78
Sul. S	0.10%	0.19	0.01	1.06
Tot. S	3.27%	1.35	0.42	6.47
SXRF	2.91%	1.24	0.54	5.40
ADL	7.70%	3.47	1.40	16.70
Mois.	9.05%	5.05	0.01	20.70
Vol.	39.70%	4.27	18.90	52.70
Fix. C	48.82%	4.95	34.60	65.40
Ash	11.44%	2.89	2.20	25.80
Btu	12,748.91	464.50	11,562.00	14,362.00
C	70.28%	3.87	55.23	80.14
H	4.95%	0.31	4.03	5.79
N	1.30%	0.22	0.78	1.84
O	8.68%	2.44	4.15	16.03
HTA	11.41%	2.95	3.28	25.85
LTA	15.28%	4.04	3.82	31.70

Abbreviations other than standard chemical symbols: organic sulfur (Org. S), pyritic sulfur (Pyr. S), sulfate sulfur (Sul. S), total sulfur (Tot. S) sulfur by X-ray flouroscence (SXRF), air dry loss (ADL), moisture (Mois.), volatile matter (Vol.), fixed carbon (Fix. C), high temperature ash (HTA), low temperature ash (LTA).

As noted above, it is difficult to predict how emitted materials will enter the environment and what roles they will play there. In this situation, it is useful to construct a simple worst-case model giving average additions to cadmium and arsenic background levels due to photovoltaics and to compare these additions with recently measured urban and suburban background levels. Annual average urban atmospheric concentrations of cadmium[4] range from below 0.01 $\mu g/m^3$ to as high as 0.036 $\mu g/m^3$, while another study indicates an average deposition rate of cadmium in residential areas of 40 $\mu g/m^2/month$.[5]

Our model for potential photovoltaic impacts due to array fires was an effectively infinite suburban neighborhood with the very high population density of 1000 residences (about 3500 people) per square mile, all powered by photovoltaic systems in which array fires occur at the rate derived above. Again assuming 10 percent release of cadmium, the average rate of release of cadmium is about 100 grams/mi^2/year or 3 $\mu g/m^3/month$. Since the neighborhood is assumed infinite, this is also the rate of deposition—less than 10 percent of mean deposition rates already occurring.

Our estimate is actually an upper bound since neighborhoods are not in fact infinite (that is, there will be areas with small or no cadmium emissions) and since not all electricity will be provided by photovoltaics. However, it should be noted that if the release fraction in fires should be greater than 0.1, the impact on net deposition would be correspondingly greater. It should also be noted that our comparison should not be taken to mean that current deposition rates are necessarily harmless: over a period of 100 years, about 50 mg of cadmium will be deposited per square meter at current deposition rates. The low solubility of cadmium compounds and their propensity for participating in ion-exchange reactions will result in a gradual buildup in soil; assuming a 10 cm mixing depth, the added cadmium will be of order 1 ppm, more than doubling present average concentrations. Since food is already a significant source of cadmium uptake (perhaps approaching half of that necessary to cause kidney damage over a lifetime), increases in deposition rates and soil concentration, which would increase concentrations in food, should be regarded with caution.

The higher release fraction for arsenic in array fires due to its high volatility implies proportionately larger contributions to arsenic backgrounds than in the parallel cadmium situation. According to the infinite neighborhood model above, the average rate of release of inorganic arsenic from residential fires would be about 300 grams/mi^2/year, or about 10 $\mu g/m^2/month$. This is also an upper bound on deposition rates; over a period of 100 years, such a deposition rate could add about 10 mg/m^2. Current deposition rates do not appear to be known. However, the fact that current ambient urban concentrations are mostly below 0.001 $\mu g/m^3$ suggest that deposition rates are significantly below those which could occur from photovoltaic sources. An impression of maximum average atmospheric concentration can be obtained by assuming a 100 m vertical mixing depth and three-day average residence time for house fire emissions. This yields an average concentration of 0.020 $\mu g/m^3$, an order of magnitude above present levels in most areas. Soil accumulation of arsenic is possible, and it is known that some food crops concentrate arsenic; its solubility means that it will also enter waterways, where it concentrates in the tissues of some edible marine species. Hazardous levels are unknown. However, it should be noted that our model gives an upper bound; it is likely that since in reality there will be large rural areas making no contribution to arsenic emissions, the contribution of fires to ambient concentrations will be lower than our estimate—perhaps below the 0.001 $\mu g/m^3$ set by limits on measurability.

Assessing the potential impacts of the larger releases associated with photovoltaic manufacturing is a more difficult task than modeling house-fire releases. Existing smelters and recovery facilities are already known to result in high local concentrations of cadmium and arsenic. Since the small particulates resulting from high-temperature processes can remain suspended for long periods (days to weeks), it is possible that these emissions will affect ambient levels far from facilities. Depending on release rates, photovoltaic array manufacturing facilities may also significantly affect concentrations and deposition rates nearby. An estimate of the possible magnitude can be obtained by assuming that emissions from array manufacturing facilities are distributed uniformly throughout an area whose electrical service is obtained from photovoltaics. In the steady state (in which an array plant with annual capacity of 200 MWe provides replacement units for 600,000 residences), the emissions from array manufacturing might be spread over about 600 mi². If releases are as high as 1 percent of throughput, this would imply an average cadmium deposition rate of 100 $\mu g/m^3$/month, far in excess of current rates. While this is a somewhat unrealistic model,* it is evident that releases must be kept far below 1 percent, and that array manufacturing facilities must be sited in a manner appropriate to the emission level actually attained. Similar cautionary statements are in order concerning inorganic arsenic emissions: a 1 percent release rate would imply an average arsenic deposition of 1 $\mu g/m^2$/month.

REFERENCES

1. Nuclear Energy Policy Study Group, *Nuclear Power Issues and Choices.* Sponsored by the Ford Foundation, administered by the MITRE Corporation. Cambridge, MA: Ballinger Publishing Co., 1977.
2. *Environmental Consideration in Future Energy Growth, Vol. I: Fuel/Energy Systems: Technical Summaries and Associated Environmental Burdens.*

---

*The actual distribution will involve very much higher concentrations near the facility and much lower concentrations farther away.

Battelle Columbus and Pacific Northwest Laboratories, Office of Research and Development, Environmental Protection Agency. Columbus, OH, 1973. As cited in *Energy Alternatives: A Comparative Analysis.* The Science and Public Policy Program, University of Oklahoma, May 1975.

3. Freudenthal, R.I., et al., *Carcinogenic Potential of Coal and Coal Conversion Products.* Columbus, OH: Battelle Columbus Laboratories, February 1975.

4. *Scientific and Technical Assessment Report on Cadmium.* U.S. Environmental Protection Agency, Office of Research and Development, No. EPA-600/6-75-003. National Environmental Research Center, Research Triangle Park, NC: March 1975.

5. Hunt, W.F., Pinkerton, C., et al., *A Study of Trace Element Pollution of Air in 77 Midwestern Cities.* Presented at 4th Annual Conference on Trace Substances in Environmental Health. University of Missouri, Columbia: 1970. As cited in [3].

Chapter 7
Indirect Impacts: Labor, Materials, and Energy

The deployment of any new technology entails a reallocation of society's productive resources—resources which have their own cost and benefit attributes. When the technology uses large quantities of these resources—which include labor, materials, and energy itself—the effects on the industrial infrastructure can be significant. There are thus indirect costs and benefits associated with changes in energy technologies. (And since the energy produced is only an intermediate good, used in many of the goods and services consumed in everyday life, increases in the resource costs of producing energy will mean that most of these goods and services will also become more expensive in social as well as economic terms.) The *costs* include health effects or other impacts resulting from the production of materials, machines, services, or capital needed to make energy-producing devices. They will also be opportunity costs—the costs of not having these resources available to pursue other socially beneficial activities. The *benefits* include having energy available to perform the functions society deems desirable and, perhaps, to have it available in ways which provide opportunities to avoid problems associated with conventional energy technologies or which allow more desirable forms of social organization (opportunity benefits rather than opportunity costs). Technological change in energy thus implies a transtion to a different spectrum of societal and individual costs, benefits, and opportunities.

The assessment and valuation of these effects is a subtle and difficult problem, requiring important social judgments as well as major efforts to define the boundaries of analysis. While social judgments must ultimately be made by society, through individual as well as collective choices, it will be useful to provide some rough measures of how substitution of photovoltaic systems for conventional sources of electricity would alter the allocation of productive resources and thus the distribution of particular indirect costs and benefits. The bookkeeping may be done along several major axes: labor, materials, energy,

70

and capital. Since capital requirements for photovoltaics are outside the scope of this analysis, we shall limit our discussion to the first three factors. Our analysis is necessarily limited by the lack of precise data concerning an immature industry.

Labor

It can be expected that technological change in the energy sector will affect the level and nature of employment in that sector. It appears, from virtually all estimates, that manufacture and use of photovoltaic systems would bring about major changes in the number and types of jobs directly involved in supplying energy. It is also probable that industries supplying materials, components, and services would be affected significantly by such a shift in electrical supply technology.

Photovoltaics is now an extremely labor-intensive technology due to the involvement of human labor in virtually all aspects of cell and array manufacture. As the technology matures, with continuous-process automation replacing handwork, there undoutedly will be a reduction in this intensity. However, apart from cell and array manufacture, most of the steps involved in transforming raw materials into electricity generating applications have already reached economically optimal labor intensity (as in raw material production) or are intrinsically labor intensive (as in transporting and installing arrays). In table 7.1 we illustrate this breakdown with a compilation of estimates for dispersed on-site silicon photovoltaic systems, using the flow sheet of chapter 5. Labor is given in millions of man-hours prorated over the electricity produced in an assumed 20-year array life. These figues are still highly uncertain and should be considered as only representative of those which might characterize a future industry. Most indirect labor requirements are not included.

The most important labor figure, and the most difficult to estimate, is that for cell and array fabrication. For silicon cells fabricated and assembled by hand, labor requirements are presently more than an order of magnitude above the high estimate shown in table 7.1. That high estimate, 10 million man-hours, is based on a partially automated plant design (20 MWe-peak of 4' by 4' modules per year) developed by the Low-Cost Silicon Solar Array project at JPL. According to projected cost data, about 60 percent of the labor is for cell manufacture. This estimate implies a labor intensity of about 40 man-hours per peak kilowatt (just for cell and array production). The extent to which reductions are possible with large-crystal silicon wafers is very uncertain.

It is possible that polycrystalline or amorphous silicon cells would provide better opportunities for applying mass-production techniques. If such cells could be made with solar-grade, as opposed to semiconductor-grade silicon, it might also be possible to reduce the relatively large labor investment

Table 7.1. Labor Requirements for Residential Silicon Photovoltaic Systems
(Millions of Man-hours/GWe-yr)(1)

	O/M	Construction
Material Procurement and Processing		
SiO$_2$ Mining(2)	0.1	
Refining(3)	4.2	
Transportation(4)	0.1 - 0.6	
Component Manufacture (cells, arrays)(5)		1 - 10
Installation(6)		1 - 4
Electrical Subsystems(7)	0.5	
Maintenance(8)	5	

(1) About 1 million 5-kw installations are required to produce 1 GWe-year of electricity per year.

(2) Currently, labor productivity in high-purity silica mining is about 0.4 tonne/man-hour; to compute total requirements we use our flow sheet (fig. 5.1 above), scaling to the GWe-yr basis using a 20-year array life and a 0.2 average capacity factor.

(3) Using present domestic production of 300 tonne/year of high purity silicon, an industry employing about 1000[1] and thus having a productivity of about 6.7 X 103 m-hr/tonne. About 500 tonnes of semiconductor Si are used for 200 MWe capacity (fig. 5.2). Note that use of solar-grade Si could reduce labor intensity per tonne but lower cell efficiency would require that more material be used.

(4) Transportation of silicon (3.5 X 104 tonne in various forms), glass, aluminum (4 X 104 tonne) or vinyl and completed panels or shingles (7 X 104 tonne) by truck for 100 to 1000 miles.

(5) This is very uncertain. About 50,000 5-kw arrays would have to be manufactured each year to sustain the 1 million installations which collectively generate 1 GWe-yr per year. The low figure above (1 million mhr) implies that it takes only 20 man-hours to perform all of the operations necessary to manufacture, prepare, and ship panels or shingles for one residential array and clearly assumes a very high degree of automation. Present silicon wafer technology is several orders of magnitude more labor intensive. See text for further discussion.

(6) Installation requirements are probably highest for shingle designs (thus compensating for reduced manufacturing labor requirements). Assuming 80 man-hours per residence, one obtains 4 X 106 man-hours/GWe-year. Of course, some of this could displace conventional roofing labor needs. Lower requirements would be associated with larger panels; however, larger labor investments might be required at the manufacturing stage. The range given here is comparable with OTA (table 7.3).

(7) Assumes 10 man-hours to integrate photovoltaic system with household, utility and (possibly) on-site electrical storage systems.

(8) Assumes an average of 5 man-hours maintenance (e.g., inspection, cleaning, repairs) per year for each residence.

(4.2 million man-hours per Gwe-year) required for refining silicon. However, it should be realized that the resulting cells would be of lower efficiency, requiring larger array areas and hence more labor in the other categories.

In order to provide a lower bound on cell and array labor requirements, we have considered a generic 5 percent efficient thin-film technology applied to float-glass produced in a facility resembling that currently used in the glass industry. The labor intensity of glass production is about nine man-hours per 100 ft^2 (for a plant producing of order 10 mi^2 of glass/year). Doubling this figure to allow for thin-film application and special handling, one obtains a labor intensity of about 4,000 man-hours per peak MWe or (prorated over 20 years) one million man-hours per GWe-year.

It is of interest to ask whether thin-film cadmium or gallium arsenide technologies would provide greater opportunities for labor reduction. While the labor intensities of purifying and compounding cadmium sulfide and gallium arsenide to solar grade are uncertain, they are undoubtedly less than that of silicon wafer technology using semiconductor grade silicon.* An informal survey of several prospective manufacturers of cadmium sulfide arrays suggests that labor intensities in pilot plants are already as low as roughly 10 million man-hours per GWe-year, even when research staff is included. There also appears to be substantial engineering conviction (based on plant design) that much lower labor intensities are possible—perhaps reaching one million man-hours per GWe-year.** Thus, industry plans for cadmium sulfide cells appear to be somewhat closer to realizing reductions in labor intensities than do those of silicon cell producers, despite lower overall photovoltaic efficiencies. However, the lower efficiencies are likely to mean that other labor requirements—at points where human involvement is essential (transportation, installation, maintenance)—will be higher. For example, if cadmium sulfide arrays have half the efficiency of silicon arrays, achieve a labor intensity in manufacture of one million man-hours per GWe-year, but involve twice as much installation and maintenance, then silicon array manufacture need only achieve a labor intensity of about five million man-hours per GWe-year in order to have comparable overall labor requirements.

When possible variations are taken into account, it appears that the total labor intensity of photovoltaic systems may be reducible to somewhere in the range of five to 25 million man-hours per GWe-year. Achieving the low end of this range would involve rather heroic technological advances and the use of relatively maintenance-free designs.***

It is instructive to compare the labor requirements for photovoltaics with those for electricity from nuclear and coal plants as shown in tables 7.2 and 7.3. There appears to be considerable variation in reported figures, perhaps due to actual variations in practice, assumptions used, or date of estimate. The latter variation may be due to regulatory requirements which have changed

---

*Large crystal gallium arsenide cells probably involve labor intensities comparable to those of silicon.

**It should be noted, however, that at present all cadmium-sulfide cell lifetimes appear to be in the three- to ten-year range, though there is optimism that 20-year lifetimes (assumed in our analysis) can be reached with the planned next generation of facilities.

***Central station and larger-scale industrial or commercial installations of photovoltaic systems would have somewhat different requirements following material procurement and processing than the residential systems considered here. Both component manufacturing and installation probably would require more labor, since subassemblies would have to be amenable to stand-alone installations and the latter would require considerable site preparation and engineering of support structures. Central station plants would also require transmission and distribution networks. However, major labor savings could result from reductions in maintenance requirements or mechanized maintenance procedures.

Table 7.2. Direct Requirements for Electricity from Coal*
(Millions of Man-hours/GWe-year)

	O/M	Construction
Fuel Cycle	1.2 - 1.9	0.1
Components	–	0.2
Power Plant	0.3 - 0.4	0.3
Transmission and Distribution	0.5	0.2
TOTAL		2.5 - 3.6

*Based on 30-year life. Data from Energy and Resources Group,[2] Office of Technology Assessment,[3] Atomic Energy Commission[4] and[5], and Project Independence Task Force Report.[6]

considerably over time. The trend appears now to be toward somewhat higher labor intensities as regulations become more restrictive and as resources deplete or must be moved greater distances, and as labor productivity decreases.

Comparison of tables 7.1 through 7.3 indicates clearly that photovoltaic systems will be much more intensive in the use of labor than coal or nuclear power, by a factor of two to five. It is at least not surprising that the difference in labor intensity between conventional and photovoltaic systems correlates well with differences in average capacity factors—a PV array produces on average at 20 percent of peak capacity while a nuclear or coal plant may operate at a capacity factor of 60 to 75 percent. Thus, even if these technologies required the same amount of human labor per unit of *peak* capacity, the average labor intensity, on a unit of energy basis, would be a factor of three or more higher. Of course, the output of photovoltaic systems in peak demand periods is also of greater value than the baseload output of coal or nuclear plants.

A gradual shift toward photovoltaic technology would help reverse the trend toward decreasing labor intensity in the electric utility component of the

Table 7.3. Direct Labor Requirements for Electricity from Nuclear Power(1)
(Millions of Man-hours/GWe-year)

	O/M	Construction
Fuel Cycle	1.2 - 1.3	0.1 - 0.2
Components	–	0.2 - 0.3
Power Plant (2)	0.3 - 0.5	0.4 - 1.0
Transmission and Distribution	0.5	0.2
TOTAL		2.9 - 4.0

(1) Based on three-year life. Data from Energy Alternatives,[7] Atomic Energy Commission[4] and[5]. Atomic Industrial Forum,[8] and Office of Technology Assessment.[3]

(2) The wide range of power-plant labor intensity appears due to regional variations—including extraordinary variations in labor productivity—and extent of regulatory changes over time and especially during construction.

energy sector. From 1961 to 1973, electric utilities increased their kilowatt output by an additional 130 percent, their revenues about 260 percent, and their construction costs about 340 percent, but employment in utilities increased only 21 percent.[9] This occurred despite increasingly restrictive regulations and decreasing labor productivity and was probably due to changes of scale in generation facilities (though economies of scale were apparently effective only in reducing unit labor requirements and not overall costs).

Photovoltaic systems would also provide a different spectrum of employment than coal and nuclear power, especially if used in decentralized point-of-use applications. Significant fractions of jobs in coal and nuclear power are in remote locations (mining, milling, transportation, and some facilities), require large but temporary worker populations in small areas (resulting in the so-called boom-town phenomenon), or are available only to highly skilled migratory workers (welders of stainless steel, constructors of containment domes or cooling towers). Fluctuations in these industries are common—due to the large scale of individual undertakings—though the overall trend is toward shortages of skilled labor in key industries (e.g., uranium and coal mining). Central station photovoltaic plants could have similar problems, though the ability to build such plants in an incremental and modular fashion could more than compensate for the magnitude of labor requirements, which are larger than for coal or nuclear plants.

In contrast, on-site photovoltaic systems would involve more workers in jobs which were as long-term and stable as those in light industries generally and could create many jobs at the community level requiring relatively modest skills (installers or maintenance personnel). While the achievement of a better congruence between employment and energy production and use may contribute little to improved economic efficiency, the local and regional social and political benefits are evident.

Whether the use of a labor-intensive energy technology like photovoltaics would also reduce unemployment—an evident social benefit—is difficult to predict. In general, the answer depends on what other shifts in the economy occur because of the change in energy supply technology. Some new jobs will obviously come at the expense of old ones: creation of maintenance jobs for PV systems may be offset partially by a reduction of jobs associated with transmission and distribution grids; photovoltaic shingle installers may displace ordinary roofers. The more subtle question concerns the extent to which the capital, material, and other demands of photovoltaics undermine the economic sources of jobs in other sectors of the economy. A simple analysis would suggest that as long as the ratio of jobs to other factors of production involved in photovoltaics is higher than the average in the economy, new jobs may result. However, relationships between the energy sector and other sectors of the economy are complex and such simple arguments must fail at some point. It is impossible to envision a healthy economy devoting all of its productive resources to the production and utilization of an intermediate good like energy. This is clearly a question deserving of further analysis.

Materials

The nature and quantities of materials required for photovoltaic systems are still very uncertain: it is not yet known, for example, whether arrays will consist primarily of vinyl shingles (in which oil is an important component), aluminum and glass sandwiches, or some other physical configuration. However, it is likely that relatively large amounts of common materials will be required for arrays producing a given amount of electricity, due to the need to cover large areas with durable assemblies. The production and fabrication of these materials will place new demands on industry and be responsible for the allocation of additional social costs to the energy sector.

A rough impression of materials requirements can be obtained by assuming that photovoltaic systems operate at 10 percent efficiency for 20 years and are fabricated from material layers 0.5 cm thick (e.g., aluminum, vinyl, glass). Prorating materials over the 20-year life (at 0.2 average capacity factor and assuming no recycle value) indicates a material allocation of $1.25 \times 10^4 m^3$ per GWe-year of electricity produced. If the material were steel, about 100,000 tonnes/GWe-year would be required; if aluminum, about 34,000 tonnes/GWe-yr; if glass, about 30,000 tonnes/GWe-year; and if vinyl, about 80,000 bbl of oil. Of course, these are extremely rough numbers; but since the assumptions involved are manifest, variations in assumptions are easily made.*

More detailed design studies for specific types of photovoltaic systems have been conducted; some of these appear to involve much larger quantities of materials. For example, a Lockheed design[10] for a 1000 MWe central station silicon PV plant (20 percent PV efficiency) was projected to require (converting to our GWe-yr basis, assuming 20-year life) 2×10^5 tonnes of steel (twice our figure above) and 6×10^4 tonnes of cement per GWe-year. The large requirements are apparently due to the tracking and concentration systems assumed. Another Lockheed study[11] projects materials requirements for cadmium sulfide array substrates. For example, the copper required for the (CU_2S) substrates, grid electrodes, and bus bar totals about 700 tonnes per GWe-year. Similar calculations can be done for other designs, presumably with wide variations in results.

It is instructive to compare these materials requirements with those of nuclear and coal plants. Assuming a capacity factor of 0.6 and a 30-year life, the major materials requirement for such conventional plants—on a GWe-year basis—are shown in table 7.4. As for photovoltaics, there are also substantial associated

---

*Interestingly, the electricity which could be obtained, at 40 percent efficiency, from 80,000 bbl of oil is about 0.006 GWe-yr. Thus, oil consumption is less than 1 percent of the electrical output of the PV arrays; note, however, that oil would have a much higher value for purposes other than electricity generation.

Table 7.4. Materials Requirements for Nuclear and Coal Plants
(Metric Tons/GWe-year) (1)

	Nuclear(2)	Coal(3)
Steel	2.0×10^3	2.6×10^3
Aluminum	1.0	14
Concrete	1×10^4	6×10^3
Copper	40	28

(1) Prorated over lifetime power output assuming 30-year life and 0.6 average capacity factor.
(2) As summarized in Krugmann.[12]
(3) Facilities Task Force Report.[13]

and antecedent materials requirements (as for rail cars for transporting coal, or machines to make pumps). The direct requirements of photovoltaics for common construction materials appear to be considerably larger than those for nuclear or coal plants. For example, the Lockheed central station design requires 100 times as much steel per GWe-year as a nuclear plant.* While such a design is probably uneconomical, it is not evident how small the materials requirements for photovoltaics could be made.

Materials requirements for photovoltaics larger than those for alternatives would mean that more of society's resources would have to be devoted to the energy sector, increasing the opportunity costs of this technology, and that the social costs associated with the production of these materials would also have to be assigned to that sector. The scale of use, relative to current output, gives some indication of the magnitude of the reallocation. If we suppose that 10 percent efficient photovoltaic panels are backed with 3 mm aluminum, for example, the annual amount required for a PV industry producing 20 GWe-peak of capacity annually in 2000 (the higher DOE goal) would be about 1.6×10^6 tonnes aluminum. This is about 45 percent of 1975 U.S. demand.**[14] The 45 percent figure given above suggests that materials-efficient photovoltaics designs should be accorded a high priority unless we are willing to commit a considerably larger fraction of our mineral resources to electricity generation than we presently do.

The special materials used in photovoltaic systems are also of interest. In table 7.5 we indicate the amounts of cadmium, arsenic, gallium, and metallurgical silicon required for an annual production of 20 GWe-peak, using the assumptions in our flow sheets of chapter 5. As an indication of the scale of these requirements, we also show U.S. domestic production and imports for

*Note, however, that the PV station supports might be used for more than 20 years.
**If one used the hypothetical Lockheed design for a 1000 MWe central station plant as the basis for providing the 20 GWe-peak capacity added annually in 2000, about 1.7×10^7 tonnes of steel would be required. This is about 15 percent of 1975 U.S. production. The PV capacity added would provide for about 0.5 percent of electricity consumption projected for the year 2000—assuming 5 percent annual growth in demand.

Table 7.5.(1)
(Metric Tons)

	Photovoltaic Requirements (tonnes/20GWe-peak)	1975 U.S. Production	1975 U.S. Imports
Cadium(2)	5,000–22,000	2,000	2,400
Arsenic(3)	550–3,600	9,600	8,300
Gallium(4)	500–3,400	6.7	6.8
Silicon (metallurgical)	140,000	430,000	46,000

(1) From[14] and flow sheets of chapter 5. All figures rounded.
(2) Cadmium thin films ranging from 3 to 8 microns in thickness, with photovoltaic efficiency ranging from 5 to 9 percent.
(3) U.S. production includes industry stocks. PV range is for wafers and thin-film cells.
(4) U.S. production includes industry stocks; U.S. demand in 1975 was about 7.5 kg. PV consumption range is for wafers and thin film cells. Note that recycling of production wastes could reduce the demand for wafer-type cells by a factor of two to three.

the most recent year for which complete data are available. While production can be expected to grow somewhat by 2000, it is evident that demand due to photovoltaics could be a substantial fraction of present production for all elements except, possibly, arsenic. For cadmium and gallium, production would have to increase dramatically to the point where appreciable dependence on foreign sources could occur. Indeed, since most of these materials are by-products of other extractive industries, the effects of greatly increased demand for byproducts on these industries should be examined. The availability of gallium is of particular concern, there being some question of whether large enough amounts could ever be available. Gallium is a trace constituent of bauxite (0.005 to 0.01 percent) and could thus be produced as a by-product of aluminum. Even the very optimistic assumption of 100 percent efficiency in recovery would imply the need to be producing 10–20 million tonnes of aluminum per year, simply to produce enough by-product gallium for 20 GWe peak capacity per year of thin film cells; present U.S. demand for aluminum is about four million tonnes.

Materials requirements are thus an important issue in choosing among energy technologies and in choosing among photovoltaic suboptions and even specific designs. In part, the choice can be made on economic grounds, since a design or particular technological option which requires excessive amounts of valuable materials will also be expensive. However, one can question whether these considerations will enter R, D and D decisions properly. Today, all photovoltaic choices are comparatively expensive and analysis of future costs are based upon projections or target goals which are sometimes not explicit about how costs attributed to various factors of production—including materials—are to be reduced. In addition, projected future economic comparisons will have to take into account the impact on basic supply industries of large-scale production of particular technologies if there is a chance that large-scale production could alter conditions in these industires. A macro approach which includes couplings between industries thus has certain advantages over the micro approach (e.g., price per peak watt) currently used to decide between alternative photovoltaic technologies. There is, for example, a danger in pursuing suboptions which appear cheapest on a unit basis if overall production can become so high as to distort the underlying supply industries, resulting in price increases in basic components.

Finally, economic calculations do not take into account important classes of social costs. The production of industrial materials and components is a major source of emissions affecting public health and the environment. When, as in the past, energy production did not place large demands on the material resources of society, these costs were borne because they were perceived as small when compared with improvements in personal and social well-being due to increased availability of goods and services. If production of materials and components must be devoted increasingly to the production of energy—which, after all, is itself an input to the production of consumable goods and services— then social costs will tend to concentrate at intermediate stages, with a smaller percentage of resources available for final goods with perceived social benefits. Thus, if photovoltaics does involve very much larger quantities of materials than current energy technologies, it is possible that the associated net social costs will offset, at least in part, the lower *direct* social costs involved in actual electricity generation. The comparative evaluation of energy technologies thus involves subtle issues of industrial structure and the demands of different energy technologies on it.

Energy

The production of a photovoltaic array, or a coal or nuclear plant and its fuel, requires an appreciable input of energy, energy which must come from pre-existing sources. For a new technology such as photovoltaics, most of this input energy must come from conventional sources—oil, coal, natural gas, and nuclear power—thus incurring an energy debt which must eventually be repaid by the new energy-producing device. Since conventional sources of energy involve social costs, ranging from health and environmental impacts to national security risks, the new technology also incurs a social cost debt. In a period of rapid growth in deployments of the new technology, it is possible that more energy from conventional sources will be used than will be produced during that period by the new devices already deployed. The result, during this growth phase, can thus be a temporary aggravation of energy supply problems and of energy-related social costs.

These observations suggest that two classes of issues need to be considered in greater detail. The first concerns the demands which a given increment of photovoltaic capacity would place on existing energy supplies and what the social consequences of this might be. The second concerns the longer-term strategy to be pursued in developing and deploying photovoltaics as a way of improving the allocation of energy supplies and social costs and benefits over time. The first issue can be characterized in specific microstructural terms (whether economic or social). The second, requiring consideration of important economic and non-economic factors of larger significance to overall social welfare, is generally beyond the scope of this study, though the inexorable need to shift to renewable energy sources suggests the wisdom of making considerable early efforts to understand and guide this transition.

To a first approximation, the energy required to produce a photovoltaic (or other energy-producing) device is usefully characterized by the time it takes that device to repay its original energy debt—the payback time. Present estimates for photovoltaic system payback times range from about two to more than ten years, depending on the cell technology assumed and on what is assumed about panel and support structures. As indicated in the above discussion, materials requirements for panel and support structures are relatively large on a unit of energy output basis, due to the need to cover large areas; the energy needed to produce these materials may account for one to three years of the total payback time, depending on design and on cell efficiency.* These energy requirements would be difficult to reduce.** Energy used to produce photovoltaic cells accounts for the remainder of the payback time. Cell production today requires large amounts of energy, due to the energy-intensiveness of particular process steps and inefficiencies in using energy-intensive materials.

Energy investments in silicon cells, using currently available large ingot wafer technology, are especially large. Reports by the Solarex Corporation (which makes such cells) indicate a payback time for cells alone of about five years.[15] However, the assumptions used introduce considerable uncertainties, some of which may lead to overestimates of payback time and some of which may lead to underestimates. For example, indirect energy investments (embodied in materials or equipment) are computed by using average relationships between dollar costs and energy investments. This procedure may err if the

---

*For example, a one GWe central station plant of the Lockheed design would require 8 X 105 tonnes of steel with an energy content of about 4 X 109 kwhe (or 12 X 109 kwht). To repay this energy debt with electricity from the photovoltaic plant would require more than two years. While energy inputs and outputs are of different quality, it is evident that some designs do not offer striking net energy advantages.

**Clearly, the easiest way to reduce energy investments in supports or panels is to use structures which would exist for other reasons as supports and to make panels serve double duty. This naturally leads to the use of roofs as supports and to the use of panels as roofing. The flexible photovoltaic shingle is perhaps the most elegant expression of this idea.

materials or equipment involved in photovoltaics (e.g., aluminum) requires more than the average amount of energy per dollar of output—a very likely possibility for items used in photovoltaic systems.[16] An independent and much lower payback projection has been made by an American Physical Society group.[17] On the expectation that energy input to cell production can be cut by a factor of three (and only 20 kwhe/m^2 are needed for panel materials and panel fabrication), the APS group projects a payback period of only about 1.2 years. The high variance in such estimates appears due to inherent uncertainties about technological possibilities. Resolution of such uncertainties—as in the case of economic uncertainties—is a key goal for research and development efforts.

The largest energy investment for current silicon cells is in the refining of metallurgical-grade to semiconductor-grade silicon, amounting to about 3.2 years payback, according to Solarex.[15] This, and the high economic cost of refining, encourage the development of solar cells using a less pure "solar-grade" silicon (and refining processes to produce such material) or more efficient ways to produce cells using semiconductor-grade silicon. The prospects for the latter appear rather limited.

Payback times for thin-film cadmium sulfide and gallium arsenide appear to be considerably smaller than for current silicon cells, though careful evaluations appear to be lacking. The crucial question to be confronted in analyses of this type is what a reasonable lower bound on energy payback might be. We believe this value to be greater than one year, simply because of the need to cover substantial areas with energy-intensive materials (aluminum, glass, copper, etc.) and perhaps less than four years. It should be noted that reducing the energy investment in silicon cells, or using low-efficiency thin films, may result in a need for larger areas of cells and arrays and thus larger overall energy investments. The analytical basis for projecting ultimate payback times deserves much more careful attention.

It is useful to compare prospective energy investments in photovoltaic systems with those of nuclear plants (which, in turn, are larger than those of coal plants). Energy investments in nuclear power production fall into two categories: those associated with the plant itself and those required for fuel. Unlike photovoltaics, energy investments in producing nuclear fuel (primarily electricity for enriching uranium) are considerable. The mining, transportation, and crushing of coal (and possibly cleaning it) involves smaller but still significant amounts of energy. However, energy costs in producing coal and nuclear fuel are incurred primarily after plant operation has begun, making for shorter lag between energy expenditure and its repayment.

In table 7.6 we show primary energy inputs for a nuclear plant and its fuel. We have separated energy investments for fuel manufactured prior to the beginning of plant operation. The initial energy investment is about 3.11×10^9 kwhe and the plant requires about 0.32×10^9 kwhe each year during its life

Table 7.6. Direct and Indirect Energy
Investments for a 1000 MWe Nuclear Plant(1)

	Kwhe	Thermal	Total(2)
	$(\times 10^9)$	$(10^{12}$ Btu$)$	$(10^9$ kwhe$)$
Plant	0.46	18.14	2.05
Fuel Prior to			
Operation(3)	0.93	1.43	1.06
	1.39	19.57	3.1
Annual Fuel	0.28	0.43	0.32
(Each year of Operation)			

(1) Includes direct energy use (as for enriching uranium) plus indirect energy involved in materials and construction of components. Derived from ERDA.[18]
(2) Thermal inputs converted to electricity at the rate of 11,400 Btu/kwhe—thus assuming that instead of being used to construct the plant the original Btus had been used to generate electricity.
(3) Assumes 10 percent of lifetime fuel requirements—first core plus first and second reloads.

for fuel. Plant output, at 60 percent average capacity factor, is 5.3×10^9 kwhe or about 5×10^9 kwhe annually when fuel production requirements are subtracted. The payback time is thus about seven months,[19] considerably less than for photovoltaic systems.

It is important to realize that the magnitude of the energy investment in a photovoltaic device does not completely characterize the role of that device even in the energy economy, let alone in a larger social context. A photovoltaic system may provide kilowatt-hours at point of use and at, or near, peak demand periods. For these or other reasons, kilowatt-hours from photovoltaics may be valued more highly than the energy used to manufacture the device, which may be produced at a remote point and off-peak. The fact that not all units of energy have the same human utility is the source of most objections to net energy analyses, especially by economists who see price as providing a more comprehensive measure of value. However, energy flow calculations can be useful in assessing a number of the impacts of technological change including those affecting long-term energy supply strategies.

One important use of energy analysis can be to help evaluate the flow of indirect social costs associated with energy and with changes in energy technology, especially when those social costs are not internalized in economic assessments or decisions. The energy invested in producing solar arrays carries with it a social cost debt. The nature of the social costs depends on the mix of energy sources used at the time the array is produced—a mix overwhelmingly dominated by fossil and nuclear fuels until the era when renewable resources provide a substantial fraction of energy supply. The magnitude of the debt can be characterized by the payback time.

For example, a photovoltaic array may require several years of operation to return the coal-generated electricity used in its manufacture. If the life of the array were only equal to the payback period, the total social cost of using coal would not be reduced; rather, the initial kilowatt-hour would be returned from a different source somewhat later in time and be accompanied by the additional social costs involved in making and using the array. However, if the lifetime of the device is greater than the payback time, and if the social costs of photovoltaics are smaller than those of coal, then the use of photovoltaics can result in a reduction of the average social costs of electricity. Thus, if the lifetime of an array is 15 years, and there are no direct social costs associated with the array, its use can reduce the average social cost of electricity by a factor of three or more, depending on the payback period. However, as we have seen, there are direct and indirect social costs involved in the production and use of photovoltaic systems. The result is that a photovoltaic device will reduce the social costs of energy appreciably only if the lifetime of the device can be made considerably greater than the energy payback time and if direct and other (nonenergy) indirect social costs can be made small.

Alleviation of the social costs associated with conventional energy sources through the use of a new energy technology is particularly difficult during the growth phase of the new technology. It is possible that deployment rates can be so high as to put a net burden on energy supplies, and thus increase the social costs due to energy, during the period of rapid growth. This occurs if the time constant characterizing the growth of deployments of the new technology is less than the energy payback time.[20] The current DOE goals, for example, call for a production capacity of 500 MWe-peak annually in 1986 and production of 10 to 20 GWe-peak per year in 2000. The time constant of photovoltaic capacity growth during the intervening years is then between 2.7 and 3.0 years. In order for photovoltaics to make a significant net contribution to energy supply, or to reduce the average social cost of energy, before the year 2000, the payback time must be considerably less than the 2.7 to 3.0 year time constant implicit in the DOE goals.

It is thus very unlikely that photovoltaics can do anything to alleviate problems of energy supply or energy-related social costs in the next two decades. The goal of deploying enough photovoltaics to make it even a relatively small component of electrical capacity in 2000[21] is incompatible with any alleviation of supply or social cost problems during this period. In fact, such near-term problems would be aggravated by photovoltaic deployments.[22] The photovoltaics program must therefore be seen as a long-term effort to change the mix of energy supply technologies. There are important reasons for doing this, but these reasons cannot include the alleviation of energy supply problems faced by this generation.

REFERENCES

1. *Mineral Facts and Problems.* 1970 ed. U.S. Department of Interior, Bureau of Mines, Bulletin No. 650. Washington: 1970. As cited in Scientific and Technical Report on Cadmium, EPA-600/6-75-003, March 1975.

2. *Evaluation of Conventional Power Systems,* Energy and Resources Group, ERG 75-7. Berkeley, CA: University of California, July 1975.

3. *Application of Solar Technology to Today's Energy Needs, Vol. 1.* Office of Technology Assessment, Congress of the United States. Washington: June 1978.

4. *Comparative Risk-Cost-Benefit Study of Alternative Sources of Electrical Energy.* Division of Reactor Research and Development, U.S. Atomic Energy Commission, WASH-1224. Washington: December 1974.

5. *Projections of Labor Requirements for Electrical Power Construction 1974-2000.* Division of Reactor Research and Development, U.S. Atomic Energy Commission, WASH-1334. Washington: 1974.

6. Clague, E., *Coal Manpower Projections, 1980.* Kramer-Associates, Inc., Washington: 1974. As cited in *Project Independence Labor Report,* Project Independence Blueprint, FEA, November 1974.

7. *Energy Alternatives: A Comparative Analysis.* The Science and Public Policy Program, University of Oklahoma: Norman, OK: May 1975.

8. *Resource Needs for Nuclear Power Growth.* Prepared by an ad hoc Froum committee within the Atomic Industrial Forum. Washington, D.C.: 1973.

9. Schipper, L., "Energy Conservation: Its Nature, Hidden Barriers, Hidden Benefits," *Energy Communications,* Vol. 2, No. 4, July 1976.

10. Sears, R.D. and Bradley, W.G., *Environmental Impact Statement for a Hypothetical 1000 MWe PV Solar Electric Power Plant.* Lockheed Missile and Space Co., Inc., and Environmental Consultants, Inc. Prepared for the Environmental Protection Agency, No. 68-02-1331. Cincinnati: August 1977.

11. Gandel, M.G., et al., *Assessment of Large-Scale Photovoltaic Materials Production.* Lockheed Missiles and Space Co., Inc. Prepared for the Environmental Protection Agency, EPA-68-02-1331. Cincinnati: August 1977.

12. Krugmann, H.H., *Capital Energy Investment in a Nuclear Power Plant.* Center for Environmental Studies, PU/CES - Report No. 37. Princeton, NJ: Princeton University, June 1976.

13. U.S. Federal Energy Administration, Project Independence Blueprint: Final Task Force Report. Facilities, VII-76. Washington, D.C.: 1974.

14. *Minerals in the U.S. Economy: Ten-year Supply-Demand Profiles for Mineral and Fuel Commodities (1966-1975).* Mineral and Materials Supply/Demand Analysis, Bureau of Mines. Washington: United States Department of Interior, 1975.

15. Lindmayer, J., et al., *Solar Breeder Energy Payback Time for Silicon Photovoltaic Systems.* Prepared by Solarex Corporation for the Jet Propulsion Laboratory, Contract No. 954606, Report No. SX/111/1Q. Rockville, MD: April 1977.

16. Solarex[15] assumes that only 2 percent of the price of any item or material is for energy—an underestimate for materials like aluminum, steel or copper and for things made from them. To compute the equivalent amount of energy an average price of 3 mills per kilowatt-hour (apparently thermal equivalent) was used, resulting in an energy/price intensity of 6.67 kwht/$. For manufactured goods, other analysts commonly use a figure of about 15 kwht/$. For comparison, it is interesting to note that sheet aluminum may be purchased for less than $2 per kilogram, but requires about 80 kwht to produce. Using a baseload power cost of 16 mills/kwhe or about 5 mills per kwht, one finds that about 20 percent of the price of sheet aluminum (or 40 percent that of bulk aluminum) is for energy. While present long-term hydroelectric contracts for power for aluminum production have lower energy prices, it is inappropriate to assume that this will be the case for new production in the long term.

17. *Principal Conclusions of the American Physical Society Study Group on Solar Photovoltaic Energy Conversion.* New York: The American Physical Society, preprint January 16, 1979.

18. *A National Plan for Energy Research, Development and Demonstration: Creating Energy Choices for the Future,* Vol. 1., U.S. Energy Research and Development Administration, ERDA 76-1. Washington: 1976. P. 113.

19. One must be careful in converting electrical and thermal inputs to a common basis. We have chosen to convert thermal inputs to electrical units as if they had been used to generate electricity (11,400 Btu being used to make 1 kwhe). It is also possible to consider how many Btus were used to make electrical inputs for the plant (at 11,400 Btu per kwhe), add thermal inputs, and to compare this Btu input with the electrical output of the plant as if it were converted to Btus (at the low efficiency, say in resistance heating, of 3412 Btu from each kwhe). This second calculation gives very different results from the first and is the source of a long-standing controversy between nuclear proponents and opponents. On the Btu basis, about 35.4×10^{12} Btu are required to start up the plant; net output (assuming a kwhe is only worth 3412 Btu) is about 14.5×10^{12} Btu annually. By this calculation then, the payback time is about 2–5 years, considerably greater than for the calculation in the text. The first calculation is the more reasonable: society should only build a nuclear plant if it is interested in having electricity and values it more highly than thermal energy; if so, the original Btus used to build the plant should be evaluated as if they had been used to generate electricity (instead of building the plant) and the output of the plant valued as electricity and not as if it were to be degraded directly into heat.

20. We use an exponential growth pattern where the capacity is $C(t) = Ae^{t/g}$, where g is the time constant for growth. The DOE goals imply a total PV capacity in 2000 of between 38 and 68 GWe-peak. The relationship between time constant and payback time is a somewhat idealized rule, valid only if energy investments are made precisely on the date the device is deployed and if no devices reach the end of their useful life during the period in question. In practice, energy investments must be made prior to

service and some devices will reach the end of their service life. The result is that the net contribution to energy supply can be negative even for growth time constants greater than the payback time.

21. Assuming a 5 percent growth in electricity production, from the present 2 trillion kwh per year, photovoltaic growth meeting the higher DOE goal would account for only 1.5 percent of year 2000 electrical output (of course, more than this output might have to go into making more PV arrays).

22. Assuming a 5 percent growth in electrical production overall, deployment of photovoltaic power systems—at the rate targeted by DOE (20 GWe capacity production in 2000) and with a 6 year energy payback time—would consume about 0.6 percent of all the electricity generated between 1986 and 2000. This is the *net* energy cost of the program: the total amount used to make PV systems minus any output from systems deployed before the end of the period.

Chapter 8
Summary and Conclusions

Our analysis indicates that there are potentially significant hazards—as well as benefits—associated with the manufacture and use of photovoltaic technologies. Direct and indirect social costs are present in all of the categories examined: occupational health, public health, and to a lesser extent, the environment.

Occupational Health

Occupational safety issues in photovoltaic technologies are likely to be qualitatively similar to those in existing industries. Inordinate safety hazards are usually relatively easy to identify; industries are accustomed to internalizing the economic costs of alleviating them. Rates of worker injury for photovoltaics are unlikely to be higher than those currently prevailing elsewhere and generally considered acceptable.

Impacts on worker health tend to be much more difficult to identify and to deal with, in this as in other industries. For all three technologies, there is the additional complication for analysis that the photovoltaic industries either do not exist yet, or are immature. Thus, quantitative comparison with coal (which results in roughly one work death for every GWe-year) is difficult. Our evaluation is necessarily limited primarily to identification of potential problem areas though a few generalizations are possible. For example, it seems clear that cadmium sulfide and gallium arsenide technologies potentially involve more serious worker health problems than silicon. Worker exposures to hazardous dusts, sprays, or vapors are possible at a number of stages in the manufacture of CdS and GaAs cells. Both cadmium and arsenic are highly toxic; arsenic is a known and cadmium a suspected carcinogen. There is a strong prospect of increasingly stringent regulatory standards for permissible workplace levels of both substances, particularly in new and potentially large industries. Large safety margins should be included in planning to minimize the chance of disruption due to regulatory change. Research and engineering emphasis on isolation of workers from these substances is desirable, though at some process steps zero exposure would be economically if not technically impractical. Care must be taken to avoid protecting worker health at the expense of public and environmental welfare (for example, by simply venting workplace emissions into the atmosphere).

The occupational health issue for silicon is considerably less problematic, though the technology is not entirely benign and the extent of worker hazard is still uncertain. The dominant risk occurs at a single process step; the refining of high-grade silica in open electric-arc furnaces. (This step would involve roughly one-fourth of the workers employed in the silicon-photovoltaics industry.) At the high temperatures attained in the furnace, vapors condensing into very small particles of crystalline silica of several forms are emitted. Inhalation can result in scarring of lung tissue (silicosis) and damage to kidneys or other sensitive organs. Though better understanding and control techniques are certainly desirable, both the proportion of potentially affected workers and the magnitude of the adverse effects seem likely to be smaller than for the other photovoltaic technologies.

In addition to the risks associated with primary materials, all three technologies present potential worker health hazards in connection with secondary substances used in cell fabrication and array assembly. Care must be taken at the stages of substrate preparation, junction formation, metallization, and encapsulation to isolate workers from potential carcinogens and to keep other exposures well below toxic levels. The synergistic effects of substances used together must be taken into account.

There is opportunity in the development of these new industries to anticipate and avoid the worker health hazards that are continuing problems in some existing industries; such consideration is to be recommended in any case, given increasingly stringent regulatory attention, especially to carcinogens. Government-sponsored research, development and demonstration programs should make these issues an integral part of their mission, avoiding the approach used in some previous technology programs of considering the problem one of "confirming" an assumed safety.

Public Health

The major direct impacts of photovoltaics on public health result from atmospheric releases of potentially hazardous materials. As with occupational exposures, the only significant risk in silicon technology appears to come from the release of particulates during the refining of silicon. For cadmium and gallium arsenide technologies, the major public risks appear to come from the release of cadmium or arsenic compounds in refining, during processing and handling in manufacture, and from fires involving arrays. For cadmium, potential health effects include kidney damage, hypertension, and, possibly, carcinogenesis; for arsenic, carcinogenesis is the primary potential effect.

It is possible to make direct comparison with the public health effects of using coal because, in addition to the well-known releases of oxides of sulfur, nitrogen, and carbon, the mining and combusion of coal also involve the emission of particulates (including silicon), cadmium, and arsenic—precisely the materials emitted in the production and use of photovoltaics.

The estimation and modeling of relevant releases suggests strongly that the direct public health impacts of photovoltaics are considerably smaller than those of coal. For both silicon and gallium arsenide technologies, the superiority of photovoltaics is relatively easy to establish. Particulate and arsenic releases from coal affect larger numbers of people at higher concentrations than emissions of these substances due to photovoltaics, even under worst-case assumptions for the latter. The comparison is somewhat more difficult in the case of cadmium since the cadmium release from coal is likely smaller than for cadmium photovoltaics. Under worst-case conditions, the collective population uptake of cadmium could be as much as ten times as large for photovoltaics as for coal. Since dose-response relationships for cadmium compounds are unknown, these exposures cannot be translated into quantitative health impacts. However, it is extremely unlikely that the adverse health effects of cadmium from photovoltaics will be as large as the combined health effects of trace elements, polycyclic hydrocarbons, and oxides of sulfur and nitrogen emitted by a coal-fired power plant.

The superiority of photovoltaics to coal in the area of public health impacts does not mean that these technologies are risk-free. Each of the technologies involves hazards—concentrations of particulates from silicon-refining arc furnaces, of cadmium oxide from materials-processing plants, and of arsenic from house fires involving arrays—that are potentially large enough to be of significant concern. Efforts to control these emissions are highly desirable, as are efforts to establish dose-response relationships for the substances involved so that the actual magnitude of public health impacts can be predicted accurately and avoided where possible.

The assessment of potential public health effects suggests several grounds for preferring silicon-based photovoltaics to the cadmium and gallium arsenide technologies, although technological change and improved epidemiological data may alter the balance. Significant potential emissions occur at only one point for silicon, but at several points for the other technologies. Silicon presents no health risk once solar arrays are installed and in use, where they will inevitably be close to—indeed, on top of—densely populated areas. Perhaps most significant in terms of the growth and development of photovoltaic technology is the fact that silicon is neither a known carcinogen (like arsenic) nor a suspected carcinogen (like cadmium). Since there is no obviously safe limit for carcinogens, the likelihood of increasingly stringent regulations governing their emissions is high; new sources will probably be the most severely restricted. Cadmium and gallium arsenide photovoltaic technologies present a series of control problems, and thus a series of opportunities for regulatory intervention and change potentially disruptive to evolving industries.

Environmental Impacts

This study examined three board categories of environmental impact: land use, thermal and climatic effects, and emissions contributing to environmental backgrounds.

Decentralized photovoltaic arrays will have no net impact on land use (except perhaps for denying roof use for other purposes). A central station solar facility would occupy an area roughly comparable to that strip mined for a coal plant with the same electrical output over its lifetime. A central station, whether solar or conventional, also requires land for power transmission lines. In addition, an uncertain amount of land will be required for the disposal of waste fly ash and slurry in the case of coal, worn-out arrays in the case of photovoltaics. For cadmium- and arsenic-based arrays, waste disposal sites would have to be remote and geologically secure against ground-water intrusion to prevent leakage of hazardous materials.

The thermal and climatic effects associated with photovoltaics are likely to be minimal. Solar panels may reduce reflectivity somewhat, raising local temperatures slightly, but this effect will be partly offset by the conversion to electricity of sunlight that would otherwise produce heat. A coal plant emits about two units of heat for every unit of electricity generated. Some of this heat is dissipated directly into the atmosphere, the rest in cooling water. Local climatic changes may result, especially if many plants are co-located. Coal emissions of particulates probably also affect global temperature, though it is still unknown whether the result is net warming or net cooling.

The most serious potential climatic effect due to coal stems from the buildup of atmospheric carbon dioxide, which now seems certain to lead eventually to global warming. Atmospheric CO_2 has increased about 10 percent since the industrial revolution and may double by the early decades of the next century. The impact of the resulting "greenhouse effect" is not yet well understood. However, a technology like photovoltaics, which permits society to avoid this risk of global climatic change, presents an obvious environmental benefit.

Photovoltaic technologies are clearly superior to coal in the area of emissions, as well. Combustion of Eastern coal in a conventional boiler with wet limestone scrubbing results in the release (per GWe-yr) of 3,000 tons of particulates, 18,000 tons of nitrous oxides, 15,000 tons of sulfur oxides, and about 400 tons of hydrocarbons. The coal required for this plant also contains dozens of heavy metals and trace elements, including cadmium (5.8 tons), lead (80 tons), nickel (48 tons), arsenic (32 tons), vanadium (75 tons), and mercury (0.5 tons). Emission rates are unknown but may be high, especially for volatile elements like mercury and arsenic.

There are no environmentally hazardous emissions produced by silicon photovoltaics. Assuming a 1 percent release rate, use of cadmium and arsenic technologies could result in the addition of 80 tons of cadmium and 7 tons of arsenic per GWe-year, respectively, to the environment. These releases could significantly increase the current environmental deposition rates of these elements; emission controls and appropriate siting are thus desirable. However, the environmental impact of coal combustion clearly exceeds that for any photovoltaic technology.

Indirect Impacts: Labor, Materials, and Energy

In its present immature state, photovoltaics is extremely labor-intensive. It appears that in a mature industry, using automation where possible, the total labor intensity of photovoltaic systems may be reducible to between 5 and 25 million man-hours per GWe-year. Coal and nuclear power require about 2.5 to 4.0 million man-hours per GWe-year. The electric utility component of the energy sector has become considerably less labor-intensive in recent decades; a shift toward photovoltaics would help reverse this trend. Such a shift would probably reduce unemployment, unless the capital, material, or other demands of photovoltaics undermine the sources of jobs in other sectors of the economy. The employment spectrum for photovoltaics may also be more desirable than that for other energy technologies, with a larger percentage of long-term stable jobs being created in existing communities.

The demand for materials for photovoltaics is also large; in this case any social impact is likely to be negative, increasing the demand and the price for some common materials. The specific kinds and amounts of materials required vary considerably with array design; it is still uncertain what materials will prove most suitable (physically and economically) for the backing and support structures onto which photovoltaic cells are mounted. In general, however, the quantities appear to be very large, at least for photovoltaic systems currently envisioned, and considerably greater than the materials requirements for coal or nuclear plants, on a unit of electrical output basis. It is obvious that materials-efficient photovoltaics designs should be accorded a high priority unless we are willing to commit a considerably larger fraction of our material resources to electricity generation than we presently do. In addition to the opportunity cost of denying these materials to other sectors, the social cost of a unit of energy will escalate as it requires larger amounts of industrial materials, each of whose production may involve its own spectrum of adverse occupational, public health, and environmental impacts.

The production of any electricity-generating facility also requires an appreciable input of energy. As with the other factors of production, photovoltaics seems likely to require more energy per unit of output than a coal or nuclear plant. Energy requirements can usefully be characterized by the time it takes a system to repay its original energy debt—the payback time. Present estimates for photovoltaic system payback times range from about two to more than ten years. The energy required to produce materials for panels and support structures accounts for about one to three years of total payback time; this requirement is difficult to reduce. Energy requirements for production of photovoltaic cells themselves are larger and more variable. Energy investments in present-technology silicon cells are especially large (payback time approximately five years). Technological improvements may reduce this requirement somewhat. Requirements for cadmium sulfide and gallium arsenide cells are uncertain but probably smaller (minimum achievable payback time probably between one and four years). The energy payback time for photovoltaics, even with optimistic assumptions, is considerably longer than the approximately seven-month payback time of a coal or nuclear plant.

A particularly important question is how the energy-intensiveness of photovoltaics affects the ability of this new technology to increase energy supply and alleviate energy-related social costs in the next few decades. In its early stages, most of the input energy for photovoltaics must come from conventional sources. In a period of rapid growth for photovoltaics, it is possible that more energy from conventional sources will be used than will be produced by the photovoltaic devices already deployed. If this occurred, the result would be a temporary aggravation of energy supply problems and of energy-related social costs. At the growth rate currently projected by DOE for photovoltaics, the energy payback time for photovoltaic systems would have to be well under two and one-half years for photovoltaics to make a net contribution to energy supply or to reduce appreciably the average social cost of energy. The photovoltaics program must therefore be seen as a long-term effort to change the mix of energy supply technologies, rather than as a means to deal with the energy problems faced by this generation.

Implications for Technological Development

The nature and magnitude of social costs are somewhat different for the three specific technologies analyzed: silicon-based photovoltaics appears to involve the least potential risk. However, for all technologies, even worst-case assumptions about releases and practices indicate a clear superiority of photovoltaics over electric power from coal, on a unit-of-power basis.*

---

*The superiority of photovoltaics to coal is not quite so clear in the area of indirect costs: if photovoltaics demanded much greater quantities of input factors of production (energy, materials, etc.) than for coal it is possible that the social costs (including oppor-

Despite the superiority of photovoltaics, the existence of serious potential risks implies a need for early attention to foresee and eliminate possible barriers to the successful use of photovoltaic systems. The greatest and most direct risk appears to result from the release of toxic or carcinogenic materials to the workplace or to the environment. For photovoltaic technologies in which this is potentially a problem—more so for cadmium or gallium arsenide than for silicon—very stringent restrictions on releases must be anticipated; government regulation is bound to become increasingly strict in the future, especially for proven carcinogens. Restrictions on occupational exposures will likely come first, followed by standards limiting releases potentially affecting the public. An example is the recent fifty-fold reduction in limits on occupational exposures to inorganic arsenic compounds on the basis of carcinogenicity. Similar actions are possible for cadmium; environmental limits are soon likely for both cadmium and inorganic arsenic. For all technologies, there is the possibility of increasingly stringent regulation of the dozens of chemical compounds used to clean, etch, dope, or otherwise treat photovoltaic cells.

These observations are extremely important to the success of photovoltaic technology programs. Unless such hazards—and the attendant regulatory regimes—are anticipated, there is serious threat to the future viability of photovoltaic options. As yet, there is not sufficient appreciation of this problem. Technology proponents, much like their predecessors in the nuclear power industry, speak of zero releases and zero risks, potentially blinding themselves to the need for attention and creating false expectations in the public mind. It is likely that the best solutions to emission problems are to be found early in the development process (perhaps even involving choices between technologies); while it may be possible to make technical fixes that reduce emissions to whatever level is required later, such fixes are likely to be expensive and could well threaten the economic viability of photovoltaics at a critical stage of development. Optimal long-term decisions about photovoltaic programs will thus require foresight about potential hazards.

Because government is so deeply involved in the research, development, demonstration, and diffusion of photovoltaic technology—and because it bears ultimate responsibility for safety and health—it is appropriate that government program decisions reflect the concerns indicated above. Functionally, one can envision two governmental roles: first, insistence on the gathering of detailed information about exposure levels in any project in which government is a partner (instrumentation, data collection, and analysis and reporting have frequently been absent in technology programs); and, second, choices between alternative development paths on the basis of facts and realistic expectations about potential hazards.

---

tunity costs) associated with these inputs could collectively exceed the direct mean that photovoltaics was far from being economically competitive and thus unlikely to be pursued on a large scale.

Beyond these important qualifications, there are reasons to be optimistic about the ability of photovoltaics to improve the balance of social costs and benefits in the energy sector. The control problems we have discussed are similar to those already being resolved in other industries and do not pose essential objections as long as enough early attention is devoted to them. The same cannot be said of the potential of CO_2 emissions from fossil fuel combustion or the risk of nuclear weapons proliferation associated with the spread of nuclear power technology, where technical or institutional remedies are difficult indeed.

Appendix A
Modeling Public Exposures

In this appendix, we first give a general description of the model used in this study to estimate the public exposures to hazardous substances due to coal combustion and various steps in the production and use of three types of photovoltaic devices. We then discuss the application of the model to these processes and give the results in terms of (1) ground level concentrations at different distances from the facility under several sets of atmospheric conditions; and (2) the number of persons exposed to each concentration level.

Release and Transport Model

As noted earlier, this study employs a simple three-dimensional Gaussian diffusion model* for the atmospheric dispersion of a general pollutant. Given the conditions of release and general atmospheric concentrations, this model gives

*The distribution used is given by Eisenbud[1]

$$X(x,y) = \frac{R}{\pi \sigma_y \sigma_z u} \; \exp \; \frac{-h^2}{2\sigma_z^2} - \frac{y^2}{2\sigma_y^2}$$

where $X(x,y)$ = groundlevel concentration (g/m^3) at point (x,y)

x,y = downwind and crosswind distances

R = release (g/sec)

$\sigma_y \sigma_z$ = crosswind and vertical plume standard deviations (both functions of x and weather stability class).

u = mean wind speed

h = effective height of release (stack height plus initial plume rise due to heat).

Unless otherwise noted, wind speed is 4 mi/hr. Additional details of calculations are given in Durand.[2]

the average concentration of material at any point downwind of the source. The conditions which must be specified include effective height of release, average wind speed, rate of release, and crosswind and vertical dispersion parameters. This model ignores plume depletion due to deposition and the effects of surface irregularities, shielding and so forth.

Typical centerline (directly downwind) ground-level distributions of material under simplified atmospheric conditions for a cold ground-level release are shown in fig. A.1. The concentration is clearly greatest near the source. Many releases, however, will occur at some elevation above ground level—due to the presence of an exhaust stack or other conditions discussed below. In this case, ground-level concentrations near the source may be very small and, depending on wind speed, maximum concentrations will occur some distance from the source. For example, at 200 meters from the source, an emission rate of 10 gm/minute and wind speed of 4 mph, ground-level concentrations are 83 mg/m^3; 12 μg/m^3; and 0.0083 μg/m^3 for release heights of 0m; 30m; and 100m respectively.

The height at which dispersion begins—the effective release height—is determined not only by the elevation of the source (a roof or stack), but also by the heat of release and by atmospheric conditions. At relatively low wind speeds and with a uniform atmospheric temperature profile (dropping about 1° C for each 100 meters of elevation) a hot release (e.g., from an arc furnace or a coal plant) will rise, expanding adiabatically and cooling, until it reaches thermal equilibrium with the atmosphere. Dispersion then occurs from this effective release height. However, if there is a high wind or a temperature inversion, the effective release height may be very different. An inversion can in effect put a lid on vertical dispersion of pollutants, greatly increasing concentrations close to the ground (dispersion cannot occur above the lid); such inversions are common in valleys, near bodies of water and at night. At higher wind speeds (on the order of 20 mph or 10 m/sec), even hot releases can be trapped in the downwind wake of buildings. This not only poses an immediate hazard due to very high concentrations, but subsequent dispersion occurs as if there had been a cold ground-level release. This may be especially important for fires affecting residential applications of some photovoltaic systems. A final condition which can affect concentrations is atmospheric instability (due, say, to surface heating by the sun) in which turbulent mixing can also result in higher concentrations (perhaps two orders of magnitude difference several kilometers away) than would be expected under idealized conditions. None of these conditions is unusual; indeed, in many areas non-ideal conditions will be the rule rather than the exception. Where possible, we take at least some of these effects into account in what follows. The assumptions involved in arriving at the presumed rates of release at each process step are discussed in the text.

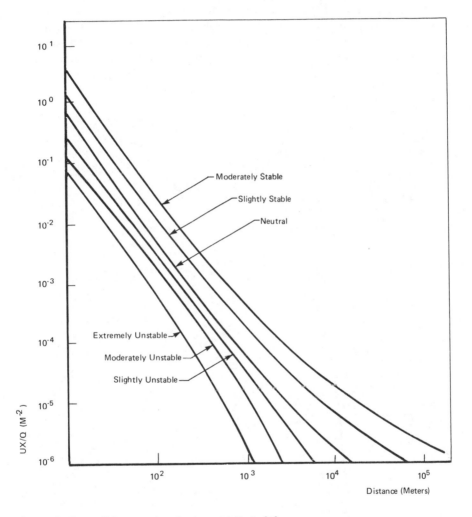

Atmospheric conditions are described more fully in [1].

X is the concentration of the pollutant in grams/m^3; U is the mean wind speed; Q is the release rate in grams/unit time.

Fig. A.1. Ground level concentrations downwind of a cold ground level release for different atmospheric conditions.

Silicon

Coal emissions generally occur (for new plants) from a 300-meter stack while silicon refining emissions may be from a stack height of 30 meters or less. Under neutral atmospheric conditions with a 4 mph wind, releases will rise about 300 meters additional for the coal plant, and 70 meters for the silicon plant, before lateral diffusion occurs. The center line ground-level concentrations of all particulates (not just submicron) downwind from both plants are shown in fig. A.2. Also shown (dashed lines) are the concentrations under

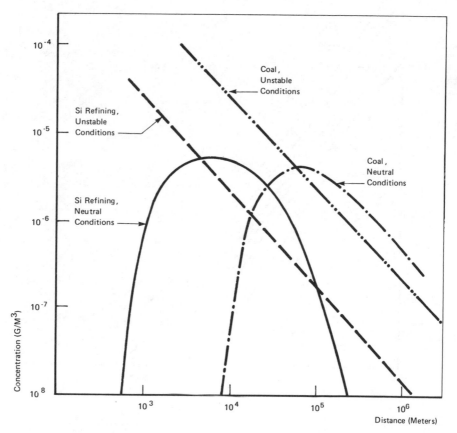

Si REFINING: particulate emissions (coke and silicon) from silicon refining; 80 g/min released; release height = 30m.

COAL: particulate emissions from coal power plant; 5000 g/min released; release height = 300m.

Fig. A.2. Centerline particulate concentrations for silicon refining and coal combustion.

unstable weather conditions, with an inversion at maximum plume height. For the unstable weather situation, coal particulate concentrations exceed those of silicon at every given distance from the plant. Under neutral conditions, coal particulate concentrations are greater at every distance beyond 10^4 meters; within a 10^4-meter radius of the plant, particulate levels are higher for silicon than for coal. In this case, the maximum concentration of silicon particulates— aproximately 5×10^{-6} g/m^3—is very similar to the maximum concentration of coal particulates. The coal maximum occurs at a greater distance from the plant, and thus over a larger area. The reason for this is that the greater height for coal releases results in significant dilution and carrying under neutral conditions, but this has much less effect under unstable conditions with an inversion. While different locations will experience different mixes of weather conditions, unstable or inversion conditions will generally prevail only a few hours a day in most areas. However, concentrations during this short period may be sufficiently high as to be the dominant source of exposure.

In order to evaluate possible effects on populations we have used this same model to compute the number of individuals exposed to different concentrations of particulates in the vicinity of each plant. Fig. A.3 shows these distributions for the weather conditions described above and for a population density of 1000 persons/mi^2.* The distributions can be adjusted for different population densities simply by scaling the vertical axis; changes in emission rates (e.g., to adjust to the fraction of particulates in the submicron range, to take into account changes in control technology, or to adjust to other plant sizes) can be accomodated by scaling the horizontal axis.

Cadmium Sulfide

Releases from cadmium recovery in zinc smelters can be modeled in a way entirely analogous to that used to describe silicon releases in the preceding section. Centerline concentrations downwind from a smelter with a capacity adequate to provide for the production of 200 MWe peak capacity per year are shown in fig. A.4. It should be noted that actual smelters may be of somewhat larger capacities (the average is about twice as large as that in our model) and may thus result in higher concentrations. It should also be noted that the 200 MWe capacity chosen as a normalization allowing comparison with coal or nuclear plant output assumes a 15-year average life, approximately three times the estimated life of present cadmium sulfide cells. If lifetimes cannot be increased substantially, the impact of cell production, per unit of

---

*The national average population density is about 70 persons/mi^2 and that of the entire eastern seaboard about 300/mi^2. However, urban area densities as high as 25,000/mi^2 are not unusual. Suburban areas generally have densities in the range 1000 − 10,000/mi^2 (0.5 − 5 persons per acre) on the average.

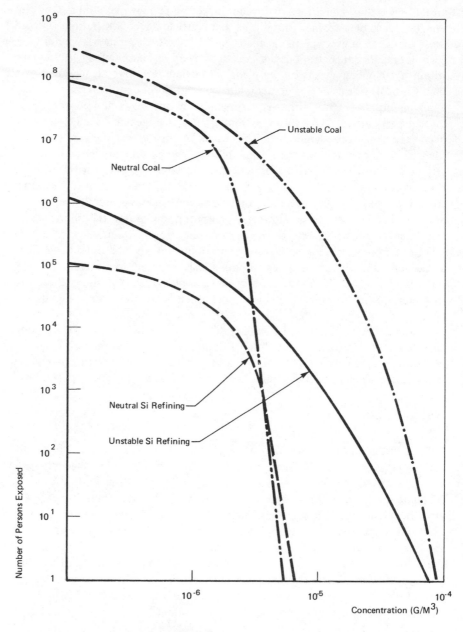

In all cases population density = 1000/mi²

Fig. A.3. Number of persons exposed to particulates from silicon refining and coal plant.

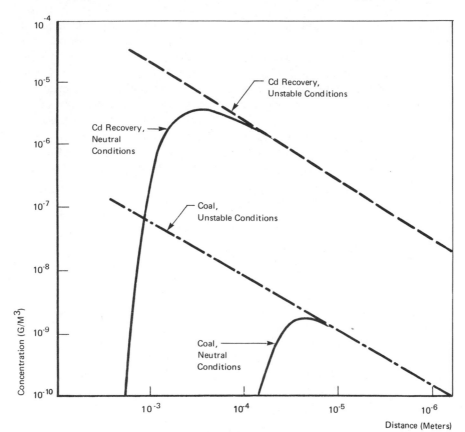

Cd RECOVERY: emissions of CdO during zinc refining, from which primary cadmium is recovered; 90 g/min. released; release height = 30 m.

COAL: emissions of CdO from coal power plant; 4 g/min. released; release height = 300 m.

Fig. A.4. Centerline cadmium concentrations: emissions from zinc smelter and coal plant.

energy ultimately delivered, will be considerably higher than computed here. Of course, greater cell life is also a condition for economic viability.

Assuming the parameters above, and a (low) population density of 100 persons/mi² (equivalent to about one person per 6 acres), it is possible to compute the number of persons exposed to given levels of concentration of cadmium at ground level under various weather conditions. With a wind speed of 4 mph, neutral conditions and an effective release height of 100 meters (e.g., 30m stack plus 70m rise due to heat of release), the 50 metric tons per year (90 g/min) release from our model cadmium recovery facility would result in the exposure levels shown in fig. A.5. Only a few people would be exposed to concentrations in excess of 1 $\mu g/m^3$; such concentrations first occur about one kilometer from the source. However, relatively large numbers of people are potentially exposed to concentrations of order 0-1 $\mu g/m^3$. Under unstable weather conditions, with an inversion at maximum plume rise, concentrations of cadmium near the smelter can be much higher, as shown in fig. A.4. In this case, however, relatively few people are affected by the higher concentration since it occurs over a small area near the smelter, as may be seen from fig. A.5.

Since weather conditions vary considerably over time, average concentrations at any given point will generally be smaller than those shown. It is interesting

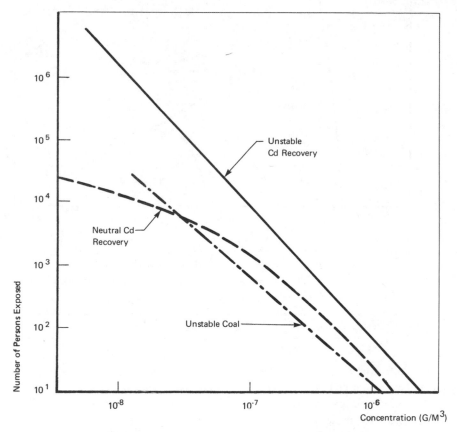

Population density around coal plant is assumed to be 1000/mi² while that for the smelter is assumed to be 100/mi². Number of persons exposed to cadmium from coal at significant concentrations under neutral conditions is too small to appear on this graph.

Fig. A.5. Number of persons exposed: cadmium releases at zinc smelter and coal plant.

to note that cadmium concentrations near existing zinc and lead smelters average 0.03 to 5 $\mu g/m^3$,[4] depending on distance and averaging period. However, population densities near existing facilities are generally low. Annual average urban concentrations of cadmium, as of 1969, ranged from less than 0.001 $\mu g/m^3$ to about 0.03 $\mu g/m^3$ for a number of cities; an exception was El Paso, Texas (which has a large lead smelter nearby) with an annual average of 0.1 $\mu g/m^3$, and a 24-hour maximum of 0.73 $\mu g/m^3$. The higher urban figures are greater than can be explained by cadmium releases from coal, which are usually well below 0.1 $\mu g/m^3$, according to our model. Low total human exposures from smelters would be due to the remoteness of such smelters from high population densities, rather than low emission rates.

As noted in the text above, it is possible to argue that cadmium emissions of zinc smelters are independent of PV use; however, this is not the case with subsequent process steps. Releases from refining and conversion of CdO to CdS appear to be a significant fraction of those involved in zinc smelting (14 tonnes versus 50 under current conditions). Centerline concentrations and numbers of persons affected at various levels are shown in figs. A.6 and A.7. Because the average population density in the area around the conversion facility (1000/mi^2) is assumed higher than that around smelters, the total human exposure due to conversion is comparable to that of smelting. Small particulates of cadmium oxide are released in both cases.

Releases during cell manufacture are more problematic: losses may occur during handling of CdS powder or in the form of CdO/CdS mixtures released following vapor deposition (contact with air of hot CdS may promote conversion to CdO). As noted in the text, plant ventilation solutions to high workplace concentrations could result in significant external public exposures. While there is no basis in current industrial experience on which to evaluate such impacts, an impression of the potential severity of the problem, and of the need for care in designing and regulating cell-manufacturing facilities, can be gained from figs. A.8 and A.9. Both assume a continuous cold ground-level release of 1 percent of the cadmium throughput. Fig. A.9 assumes a population density of 1000/mi^2. Unless measures to reduce, dissipate, or isolate emissions are taken, large numbers of people could be exposed routinely concentrations of cadmium compounds exceeding 1 $\mu g/m^3$. Measures to reduce concentrations in populated areas include basic confinement (including closed process lines), high stacks or heating of exhausts, and remote siting.

The final mode of public impact from cadmium is through fires in arrays. Assuming an instantaneous release of 10 percent of the cadmium in a 5 kw residential array, representative exposures downwind (centerline) of a single house fire under various weather conditions can be projected, as shown in fig. A.10. These are maximum values and, in general, individual will receive smaller doses due to shifting wind patterns. Unlike the continuous exposures due to smelters and manufacturing facilities, fires are episodic, with concen-

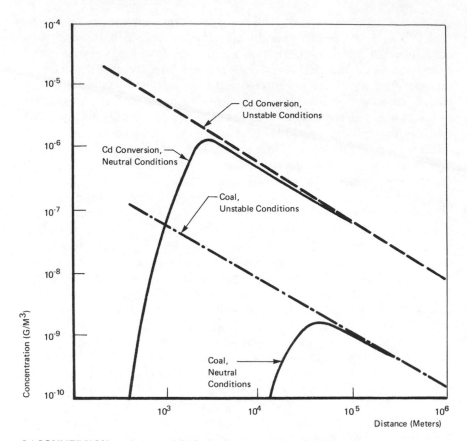

Cd CONVERSION: emissions of CdO during the refining of cadmium and its conversion to cadmium sulfide; 30 g/min. released; release height = 30 m.

COAL: emissions of CdO from coal power plant; 4 g/min. released; release height = 300 m.

Fig. A.6. Centerline concentrations of cadmium: cadmium refining and conversion facility and coal plant.

trations of cadmium rising rapidly and then falling off as releases dissipate. The exposures shown in fig. A.10 are thus sums over concentrations multiplied by the time which each concentration persists. The result is a time-integrated exposure, expressed in gram-minutes/m³. For example, exposure to 1 $\mu g/m^3$ for one hour would give a total exposure of 60 $\mu g\text{-min}/m^3$. To convert this to an individual dose, one multiplies by a breathing rate (in m^3/minute) and by a bodily retention factor.

The exposures in fig. A.10 must thus be multiplied by 7 to 15 $\times$ $10^{-3} m^3/$ minute and be a retention factor of 0.1 to 0.4 to obtain individual doses. The

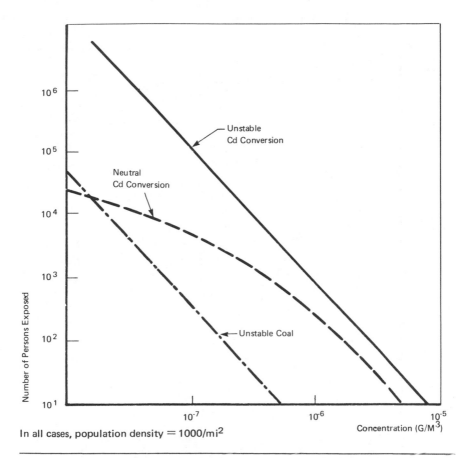

In all cases, population density = 1000/mi²

Fig. A.7. Number of persons exposed: CD refining and conversion plant and coal plant.

largest exposures occur under unstable or downwash conditions and range from 10^{-5} to 10^{-3} g-min/m³. The corresponding individual doses could thus be as high as 6 μg retained. Only a few individuals would experience such doses. Indeed, such an individual exposure would result only under downwash conditions during a fire affecting a neighboring house upwind. The probability of a given individual experiencing such an exposure is only of order 10^{-4} per annum, assuming that downwash conditions occur about 10 percent of the time. The number of people experiencing each exposure is shown in fig. A.11,* assuming a residential population density of 5000/mi²—about that of Los Angeles or many residential suburbs.

_____

*The maximum dose resulting from each exposure—assuming 40 percent retention and a breathing rate of 20 m³/day—can be obtained by multiplying the exposure by 5 6 X 10⁵.

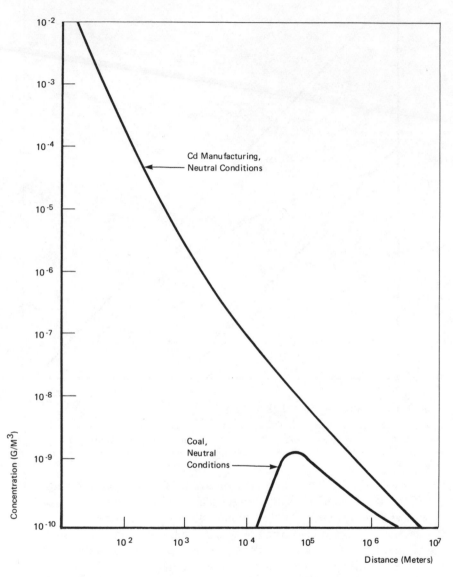

Cd MANUFACTURING: emission of CdS during the manufacture of cadmium photovoltaic cells; 3.8 g/min. released; release height = 0

COAL: emissions of CdO from coal power plant; 4 g/min. released;
release height = 300 m.

Fig. A.8. Centerline concentrations: cadmium from manufacturing facility and coal plant.

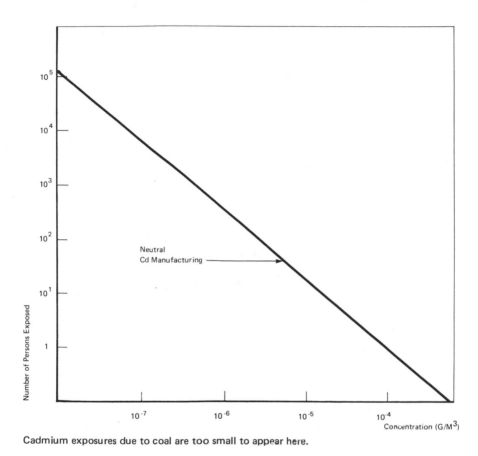

Cadmium exposures due to coal are too small to appear here.

Fig. A.9. Number of persons exposed: cadmium releases from manufacturing facility.

To compute impacts which can be compared directly with our coal plant, it is necessary to add up the effects of the 200 fires which might occur annually in the 600,000 residential installations providing as much total electricity as the coal plant. To do this, one simply multiplies the numbers of people exposed to each concentration by 200.

Gallium Arsenide

For purposes of comparison with subsequent process steps, the ambient center-line exposure levels associated with a (hypothetical) smelter/production plant releasing 5 percent of arsenic throughput and producing enough arsenic for 200 MWe of capacity are shown in fig. A.12. In principle, a reduction of about

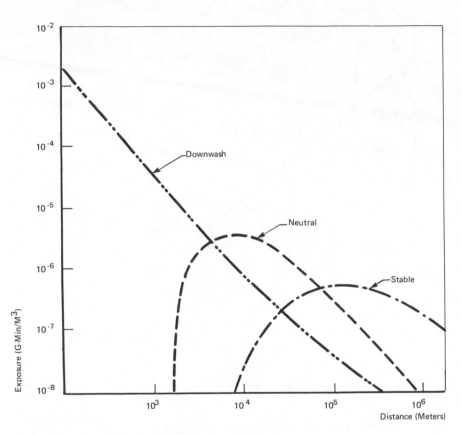

House fire releasing 200 g of CdO (10% release fraction)
Wind speed = 4 mph (20 for downwash)
Release height = 20 m (ground-level for downwash case)

Fig. A.10. Centerline exposure due to cadmium emission from a single house fire.

a factor of ten in this release rate may be possible, though lack of information on how arsenic emissions actually occur (some arsenic may be emitted as a vapor which could only be controlled by providing appropriate adsorptive surfaces) does not allow precision in such estimates. Also shown are exposure levels due to arsenic releases from coal combustion. Fig. A.13 shows the number of persons exposed, as a function of concentration, for a uniform population density of 100 persons/mi² near the smelter. For smelters located near higher populations, the curve should be displaced to the right. Also shown is the population exposure due to a coal plant, assuming 1000 persons/mi². Coal plant emission effects greatly exceed those of smelters except in the immediate vicinity of the smelter.

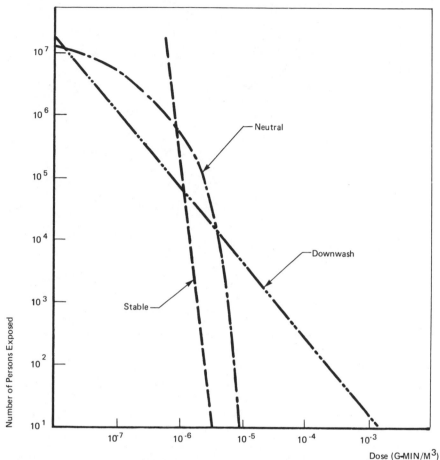

House fire releasing 200 g of CdO
Population density $= 5000/mi^2$
Release height $= 20$ m (ground level for downwash case)
Wind speed $= 4$ mph (20 for downwash)

Fig. A.11. Number of persons exposed to cadmium from a single house fire.

For the As reduction and purification step, centerline exposure levels under various weather conditions are those shown in fig. A.14* for an emission rate of 1 percent throughput. Total population exposure levels are shown in fig. A.15. No individuals appear to be exposed to concentrations exceeding 1 $\mu g/m^3$.

The concentrations and public exposures associated with the gallium arsenide compounding step will be similar to those of reduction and purification as

*We assume a 30 m stack.

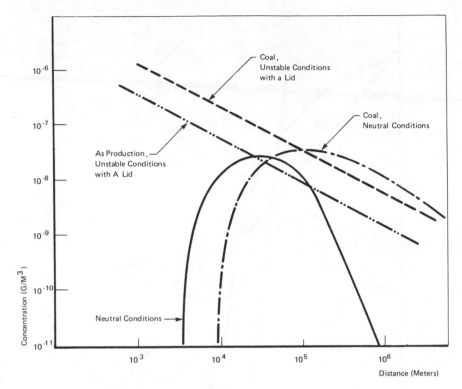

As PRODUCTION: Emissions of As_2O_3 during the recovery of arsenic trioxide from ores and flue dusts; 6 g/min. released; release height = 30 m.

COAL: Emissions of As_2O_3 from coal power plant; 30 g/min. released; release height = 300 m.

Fig. A.12. Centerline arsenic concentrations: emissions from arsenic production and coal combustion.

shown in figs. A.14 and A.15. However, the emissions from cell manufacture and array assembly will present a different dispersion picture, since temperature and height of release (as well as the nature of the potential emission itself) will probably be unlike those prevailing at earlier production steps. If 1 percent of arsenic throughput is lost to the atmosphere at this manufacturing stage, the concentrations shown in fig. A.16 could occur. Within one kilometer of the plant, concentrations range from 1 to 100 $\mu g/m^3$. The numbers of individuals exposed at each level are shown in fig. A.17 for a population density of 1000 persons/mi².

As with cadmium, the final mode of public impact from gallium arsenide technology is through fires in arrays. In fig. A.18 we show total exposures

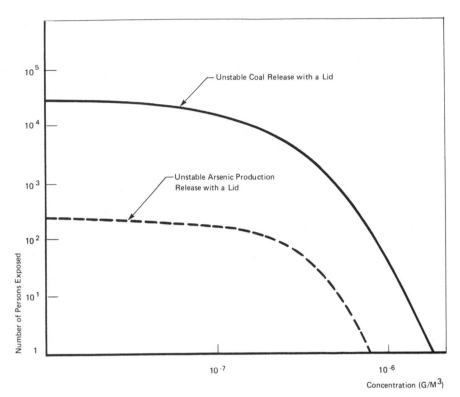

Population density around coal plant is assumed to be 1000/mi^2 while that for the arsenic production facility is assumed to be 100/mi^2.

Fig. A.13. Number of persons exposed: arsenic emissions from arsenic production and coal plant.

as a function of distance from a fire resulting in the release of 70 percent of the arsenic (as the trioxide) in a residential array, under three different weather conditions. Total exposures are in gram-minutes per cubic meter—a quantity which may be thought of as an exposure to a certain amount of arsenic per cubic meter for some period of time (e.g., 100 μg-minutes/m^3 can be thought of as a ten-minute exposure to 10 μg/m^3). To obtain a human dose, one must multiply by a breathing rate (e.g., 10-20 m^3/day and a bodily retention factor (e.g., 10 percent). In fig. A.19 we show the total number of persons receiving each exposure, assuming a population density of 5,000/mi^2.

It is apparent from fig. A.19 that individual exposures on the order of 100 μg-min/m^3 could occur for significant numbers of people. With a breathing rate of 10 m^3/day, a person exposed to 100 μg-min/m^3 would inhale about

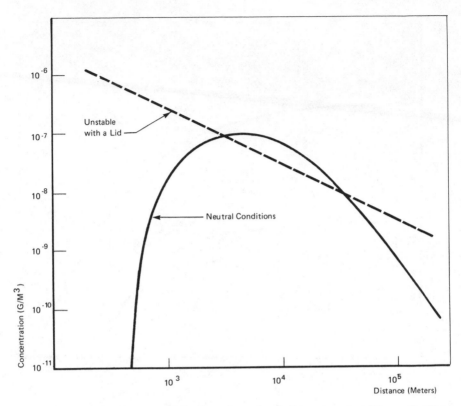

Emissions of As_2O_3 during the reduction of arsenic trioxide to metal and its purification;
1 g/min. released; release height $= 30m$.

Emissions of As_2O_3 due to coal combustion are shown on Figure A-12

Fig. A.14. Centerline arsenic concentrations: arsenic reduction and purification facility.

10^{-7} grams of arsenic; the retention factor does not appear to be known. It
is unclear whether this results in any significant individual cancer risk. We
have summed up the individual exposures and find, according to our simple
model, that the total amount of arsenic inhaled by people per house fire ranges
from 0.01 gram to 0.065 gram. These are probably upper bounds.* To compare with a coal plant, for which about 30 grams of arsenic are inhaled per
year (see fig. A.13), it is necessary to multiply by the number of fires occurring

---

*Roughly one million persons are involved in the summation, based on the population
density assumed. Presumably, this overstates the population actually involved. However,
it should be noted, we have neglected long-term suspension and resuspension of small
particulates.

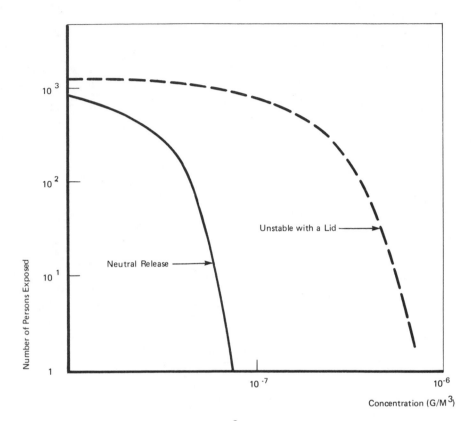

Population density is assumed to be 1000/mi²

Fig. A.15. Number of persons exposed: arsenic reduction and purification facility.

in arrays giving the same electrical output per year as the coal plant. With an estimated 200 fires per year, the total annual inhalation due to array fires ranges from about 10 to 50 percent that of coal combustion for the same amount of electricity generated.

It is also of interest to ask how array fires might alter existing average ambient concentrations of arsenic. For this purpose, we assume a universal neighborhood with a uniform population density of 5000 persons/mi².* Array fire incidence is then about ½ per mi² per year. Using the groundlevel

---

*As above, the universal neighborhood model is a simple way to account for effects which may otherwise be difficult to bookkeep—the simplification comes from the fact that any airborne material leaving any limited region is compensated for exactly by an inflow of material from similar regions nearby.

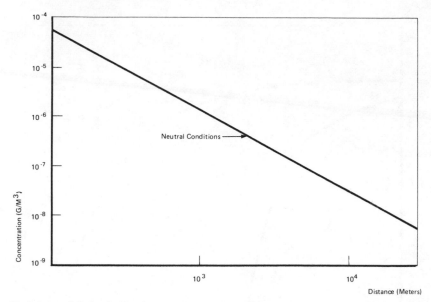

Emissions of GaAs during the manufacture of gallium arsenide photovoltaic cells; 1 g/min.
released; release height = 0

Fig. A.16. Centerline arsenic concentrations: manufacturing facility.

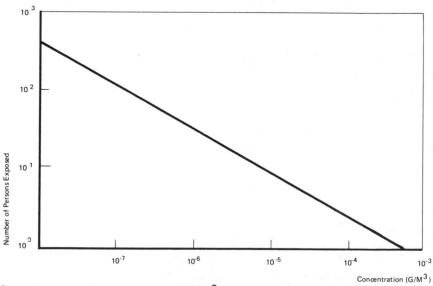

Population density is assumed to be 1000/mi^2

Fig. A.17. Number of persons exposed: arsenic releases from manufacturing facility.

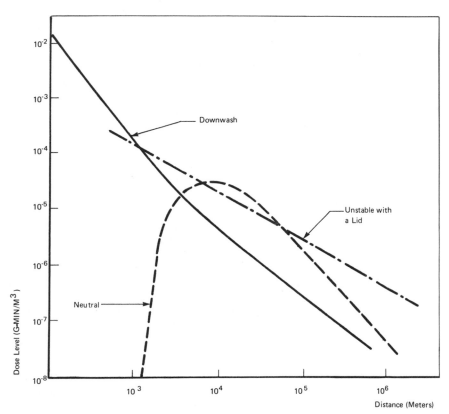

House fire releasing 900 g of As$_2$O$_3$ (70% release fraction)
Wind speed = 4 mph (20 for downwash)
Release height = 20 m (ground-level for downwash)

Fig. A.18. Centerline exposure due to arsenic emission from a single house fire.

population exposures in fig. A.19, the average individual amount of arsenic inhaled per year is about 1-6 μg (the range being due to weather conditions). This is equivalent to exposure to a continuous ambient concentration of about 0.0003 to 0.002 μg/m^3—values which could be compared with prevailing urban concentrations (most below the limit of detection of 0.001 μg/m^3). Thus, fires in gallium arsenide arrays would not grossly affect average ambient levels as presently measured, except under adverse weather conditions.

REFERENCES

1. Eisenbud, M., *Environmental Radioactivity,* 2nd edition. New York: Academic Press, 1973.

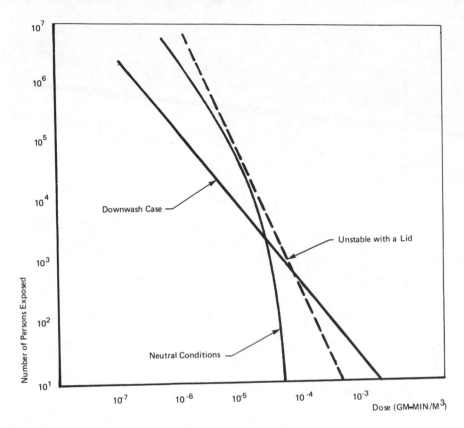

House fire releasing 900 g of As_2O_3 (70% release fraction)
Wind speed $= 4$ mph (20 for downwash)
Height of release $= 20$m (ground-level for downwash)
Population density $= 5000/mi^2$

Fig. A.19. Number of persons exposed to arsenic from a single house fire.

2. Durand, J., *Policy Implications Derived from Assessment of Health and Environmental Costs of Photovoltaic Power.* Master of Science Thesis, Massachusetts Institute of Technology, June 1978.
3. The population distributions shown in fig. A.3, and in subsequent figures in this appendix, are smooth curves drawn through histograms showing the number of people exposed to concentrations in certain ranges. The intervals chosen were 2.5×10^{-a} to 7.5×10^{-a} and 7.5×10^{-a} to $2.5 \times 10^{-a+1}$; it is necessary to specify these intervals in order to interpret properly the population figures shown. The populations were determined by computer analysis in which the area around a facility was broken down into a fine grid; the computer was then asked how many grid points fell

into each of the ranges above; finally, the number of grid points in each range was multiplied by the number of people associated with a single grid unit.

4. *Scientific and Technical Assessment Report on Cadmium.* U.S. Environmental Protection Agency, Office of Research and Development, No. EPA-600/6-75-003. National Environmental Research Center, Research Triangle Park, NC, March 1975.

Index

About the Author

Thomas L. Neff (doctorate in physics, Stanford) is Principal Research Scientist and Manager of the International Energy Studies Program at the Energy Laboratory of the Massachusetts Institute of Technology. He has served as advisor to the U.S. Department of State, the U.S. Arms Control and Disarmament Agency, the Office of Science and Technology Policy and to other public and private organizations. Dr. Neff's current research interests include international markets in oil and uranium and the problems of the transition from depletable to renewable energy resources.